Science: a third level course

S343 INORGANIC CHEMISTRY

Block 6 **ORGANOMETALLIC CHEMISTRY**

The Open University

Prepared by an Open University Course Team

S343 Course Team

Course team chairs	*Stuart Bennett, Elaine Moore, Lesley Smart*
Authors	Block 1 *David Johnson*
	Block 2 *Rob Janes and Elaine Moore*
	Block 3 *Kiki Warr, with contributions from David Johnson*
	Block 4 *Stuart Bennett*
	Block 5 *Michael Mortimer*
	Block 6 *Ivan Parkin, with contributions from Dr M. Kilner, Professor K. Wade* (University of Durham) *and F. R. Hartley* (Royal Military College of Science, Shrivenham)
	Block 7 *Paul Walton* (University of York), *with contributions from Lesley Smart*
	Block 8 *Lesley Smart, with contributions from David Johnson, Kiki Warr and Elaine Moore*
	Block 9 *David Johnson*
Consultants	*Dr P. Baker* (University College of North Wales)
	Dr R. Murray (Trent Polytechnic)
Course managers	*Peter Fearnley*
	George Ford
	Wendy Selina
	Charlotte Sweeney
Editors	*Ian Nuttall*
	David Tillotson
BBC	*Andrew Crilly*
	David Jackson
	Jack Koumi
	Michael Peet
Graphic artists	*Steve Best*
	Janis Gilbert
	Pam Owen
	Andrew Whitehead
Graphic designers	*Josephine Cotter*
	Sarah Hofton
	Jane Sheppard
Assistance was also received from the following people:	*George Loveday* (Staff Tutor)
	Joan Mason
	Jane Nelson (Staff Tutor)
Course assessor	*Professor J. F. Nixon* (University of Sussex)

This publication forms part of the Open University course S343 Inorganic Chemistry. The complete list of texts which make up this course can be found at the back. Details of this and other Open University courses can be obtained from the Course Information and Advice Centre, PO Box 724, The Open University, Milton Keynes MK7 6ZS, United Kingdom: tel. +44 (0)1908 653231, e-mail general-enquiries@open.ac.uk

Alternatively, you may visit the Open University website at http://www.open.ac.uk where you can learn more about the wide range of courses and packs offered at all levels by The Open University.

To purchase a selection of Open University course materials visit the webshop at www.ouw.co.uk, or contact Open University Worldwide, Michael Young Building, Walton Hall, Milton Keynes MK7 6AA, United Kingdom for a brochure. tel. +44 (0)1908 858785; fax +44 (0)1908 858787; e-mail ouwenq@open.ac.uk

The Open University
Walton Hall, Milton Keynes MK7 6AA

First published 1989. Second edition 1994. Reprinted with corrections 1999, 2004.

Copyright © 1989, 1994 The Open University

All rights reserved. No part of this publication may be reproduced, stored in a retrieval system, transmitted or utilised in any form or by any means, electronic, mechanical, photocopying, recording or otherwise, without written permission from the publisher or a licence from the Copyright Licensing Agency Ltd. Details of such licences (for reprographic reproduction) may be obtained from the Copyright Licensing Agency Ltd of 90 Tottenham Court Road, London W1T 4LP.

Open University course materials may also be made available in electronic formats for use by students of the University. All rights, including copyright and related rights and database rights, in electronic course materials and their contents are owned by or licensed to The Open University, or otherwise used by The Open University as permitted by applicable law.

Except as permitted above you undertake not to copy, store in any medium (including electronic storage or use in a website), distribute, transmit or re-transmit, broadcast, modify or show in public such electronic materials in whole or in part without the prior written consent of The Open University or in accordance with the Copyright, Designs and Patents Act 1988.

Edited, designed and typeset by The Open University.

Printed and bound in the United Kingdom by Henry Ling Limited, The Dorset Press, Dorchester.

ISBN 0 7492 5089 5

s343block6i2.3

STUDY GUIDE FOR BLOCK 6		5

1	**INTRODUCTION**		6

2	**STRUCTURES AND BONDING OF ORGANOMETALLIC COMPOUNDS**		6
	2.1	Ligand types	6
	2.2	Structural types and the Periodic Table	10
	2.3	Bonding in organometallic compounds	10
	2.4	Summary of Section 2	12

3	**STABILITY AND REACTIVITY OF *MONOHAPTO* ORGANOMETALLIC DERIVATIVES**		12
	3.1	Thermal stability	12
	3.2	Stability to oxidation	14
	3.3	Stability to hydrolysis	15
	3.4	General reactivity	15
	3.5	Summary of Section 3	16

4	**METHODS OF FORMING METAL–CARBON BONDS**		17
	4.1	Reaction of a metal with a haloalkane — metallation	17
	4.2	Reaction of a metal with an organo derivative of another metal — transmetallation	18
	4.3	Reaction of a metal halide with an organo derivative of another metal — metathetical reactions	18
	4.4	Insertion of alkenes or alkynes into metal–hydrogen bonds — hydrometallation	19
	4.5	Summary of Section 4	20

5	**SYSTEMS CONTAINING *MONOHAPTO* LIGANDS**		20
	5.1	Organomagnesium compounds	21
	5.2	Organolithium compounds	23
	5.3	Organoboron and organoaluminium compounds	24
	5.4	Organo derivatives of silicon and the heavier Group IV elements	25
	5.5	Organo derivatives of Group V and Group VI elements	28
	5.6	*Monohapto* derivatives of the transition metals	31
	5.7	Stability of *monohapto* organometallic compounds	33
	5.8	Metal carbenes and carbynes	35
	5.9	Agostic interactions	38
	5.10	Summary of Section 5	38

6	**ACYCLIC *POLYHAPTO* SYSTEMS**		39
	6.1	Introduction	39
	6.2	Bonding of acyclic *polyhapto* ligands	40
		6.2.1 Bonding of *dihapto* (η^2) systems to metals	40
		6.2.2 Bonding in other acyclic *polyhapto* derivatives	43
	6.3	The *polyhapto* classification of organic ligands	46
	6.4	'Bucky balls'	47
	6.5	The eighteen-electron rule	48
	6.6	Isolobal analogy	50
	6.7	Wade–Mingos rules	54
	6.8	The chemistry of alkene complexes	58
		6.8.1 Synthetic aspects	59
		6.8.2 Reactions with electrophiles	60
		6.8.3 Substitution reactions	60
		6.8.4 Nucleophilic addition to polyenes — Green's rules	60
	6.9	Summary of Section 6	63

7	**CYCLIC *POLYHAPTO* SYSTEMS**		**64**
	7.1	Structure and bonding in η^5-cyclopentadienyl complexes	64
	7.2	Other cyclic *polyhapto* ligands	67
	7.3	Range of derivatives of cyclic *polyhapto* ligands	69
	7.4	The chemistry of metallocenes	70
		7.4.1 Synthetic aspects	71
		7.4.2 Physical aspects	72
		7.4.3 General properties and reactions	72
	7.5	Summary of Section 7	73

8	**CATALYSIS BY TRANSITION-METAL COMPLEXES**		**74**
	8.1	Alkene metathesis	75
	8.2	Heterogeneous catalysis: the Zieglar–Natta process	76
	8.3	Homogeneous hydrogenation of alkenes: Wilkinson's catalyst	77
	8.4	Fischer–Tropsch catalysis	82
	8.5	Essential steps in catalytic cycles involving transition-metal complexes	83
		8.5.1 Vacant sites	83
		8.5.2 Oxidative addition and reductive elimination	84
		8.5.3 Ligand combination reactions	85
		8.5.4 Attack on a coordinated ligand	87
		8.5.5 β-Hydrogen abstraction	87
	8.6	Properties of metals in homogeneous catalysis	88
	8.7	Properties of ligands in homogeneous catalysis	89
		8.7.1 Electronic properties of the ligand	89
		8.7.2 Charge on the ligand	89
		8.7.3 Steric effects	89
	8.8	Summary of Section 8	89

9	**SUMMARY OF BLOCK 6**	**90**

OBJECTIVES FOR BLOCK 6	**90**

SAQ ANSWERS AND COMMENTS	**93**

STUDY GUIDE FOR BLOCK 6

This Block represents the beginning of the second half of the Course. You have already been introduced to the major theories of chemistry which lie at the heart of studies in transition metals, and in the remainder of the Course we shall explore four major, rapidly developing areas of inorganic chemistry. Organometallic chemistry is the first of these, and is covered in this Block.

Each of the remaining four Blocks (6–9) of the Course is equivalent to two Units, and is designed to occupy two study weeks. You should aim to reach the end of Section 5 of Block 6 by the end of the first week. This week should also include your study of the videocassette programme on silicones. Fuller details of the contents of the videocassette are to be found in the S343 *Audiovision Booklet*. You will find that it is useful to have ready access to your model kit when you study the Block, so that you can try to explore the geometry of the many new compounds that you will meet. It will also be advantageous to have the S343 *Data Book* handy, since it gives some guidance on the nomenclature of coordination compounds and most of the ligands you will meet in this Block.

Organometallic chemistry is an extremely broad area of chemistry, to which many reactions apply and underlain by a number of principles. You will not be expected to remember all the individual reactions, but after a detailed study of this Block you should be able to discuss the following points.

- Predict which fragments of a compound are isolobal, for example CR_2, $Fe(CO)_4$, NH.

- Account for the bonding in σ-bonded compounds such as $[W(CH_3)_6]$, and the π-bonding in compounds such as $[PtCl_3(\eta^2\text{-}C_2H_4)]^-$, η^4-butadiene and η^4-cyclobutadiene compounds, and ferrocene $[Fe(C_5H_5)_2]$.

- Determine whether organometallic complexes such as $[Mn(CO)_5Me]$, $[Fe(\eta^4\text{-}C_4H_6)(CO)_3]$, $[Cr(\eta^6\text{-}C_6H_6)_2]$, $[Fe(\eta^5\text{-}C_5H_5)_2]$ obey the eighteen-electron rule.

- Predict the sites of nucleophilic addition on cationic polyenes such as $[Fe(C_5H_5)_2]^+$ and $[W(\eta^5\text{-}C_5H_5)(\eta^2\text{-}C_2H_4)(CH_3)]^+$ by using Green's rules.

- Formulate the structures of organometallic clusters like $[Os_3(CO)_{12}]$ and $[Co_4(CO)_{12}]$ using the Wade–Mingos rules.

- Account for the decomposition pathways of organometallic compounds, especially β-hydride elimination, orthometallation and α-H abstraction.

- Differentiate between kinetic, thermodynamic, oxidative and hydrolytic labilities of various organometallic compounds.

- Account for all the main stages in a catalytic cycle, and describe what happens at each step—for example, oxidative addition, reductive elimination, insertion, dissociation, ligand combination and rearrangements.

- Detail the main synthetic routes to organometallic complexes such as BuLi, PhMgBr, BEt_3, $GeMe_4$, $[Fe(\eta^5\text{-}C_5H_5)_2]$ and $[V(\eta^5\text{-}C_5H_5)_2]$, including the terms 'metallation', 'hydrometallation', 'metathesis' and 'transmetallation'.

- Describe the double bond rule and give examples of where it has limitations.

You may find it helpful to write these various points down on paper and refer to them as you read the Block. At this stage you are not expected to understand what they mean, but you should recognise that they are the most important parts of the Block and they should serve to focus your study.

1 INTRODUCTION

Block 6 is concerned with organometallic chemistry. This field of chemistry is intermediate between organic chemistry and traditional inorganic chemistry, and encompasses all compounds in which there is a carbon-to-metal or carbon-to-semi-metal bond.

Organometallic catalysts are important industrially because they have a pivotal role in the manufacture of a number of carbon-containing compounds. The number and variety of organometallic complexes is limited almost only by the imagination.

In this Block we classify organometallic compounds by the *hapto* convention, wherein the *hapticity* of a ligand is defined as the number of carbon atoms directly bonded to the metal. *Monohapto* organo ligands (linked through one carbon atom) predominate for the main-Group organometallic compounds. On the other hand, organometallic complexes containing *polyhapto* ligands, where more than one carbon atom of a ligand is bound to a metal, are more common for transition metals, and often show a wide variety in structure and bonding.

We shall also discuss the factors that stabilise organometallic complexes, use rules of electron counting to rationalise the bonding in organometallic clusters, and give insights into the reaction chemistry of this diverse class of compounds.

In the final Section a number of reactions catalysed by organometallic compounds are briefly discussed. Different reactions such as alkene hydrogenation, polymerisation and oxidation will be presented to typify the diverse science which comprises organometallic chemistry.

2 STRUCTURES AND BONDING OF ORGANOMETALLIC COMPOUNDS

2.1 Ligand types

Before we consider the wide variety in structure, bonding and reactivity within organometallic chemistry, it is important to have a working definition of what an organometallic compound is. A compound can be defined to be organometallic if the M—C bond is polarised $M^{\delta+}-C^{\delta-}$; this means that the M atom must be less electronegative than carbon. Therefore the halogens, nitrogen and oxygen are not considered to form organometallic complexes, because for these elements the polarisation is $M^{\delta-}-C^{\delta+}$. However, this leaves some eighty elements that can form organometallic compounds.

One traditional way of evaluating the bonding in organometallic compounds is to consider the idea of bonding modes. This should be decided by classifying the ligand into one of the following three types:

(a) one that is attached through only one of its carbon atoms to a single metal atom;

(b) one that is attached through two or more of its carbon atoms to one metal atom;

(c) one in which the organic ligand may function as a bridge linking two or more metal atoms.

Figures 1–3 illustrate some of the diversity in organometallic chemistry. These Figures include a number of ligands with which you will be unfamiliar; the chemistry of these complexes will be considered later in the Block.

The ligands shown in the complexes in Figure 1 are virtually all *monohapto* ligands, because they bond to the metal through only one carbon atom of a particular ligand. If two or more atoms of the ligand are directly bonded to the same metal atom, as in the examples shown in Figure 2(a–c), the ligand is said to be *polyhapto,* and this can be indicated by the prefix *dihapto-*, *trihapto-*, etc. In the formula, the **hapticity** of a ligand — that is, the number of ligand atoms directly bonded to the metal atom — is indicated by the Greek letter η (eta), followed by a superscript number to indicate the *hapticity*.

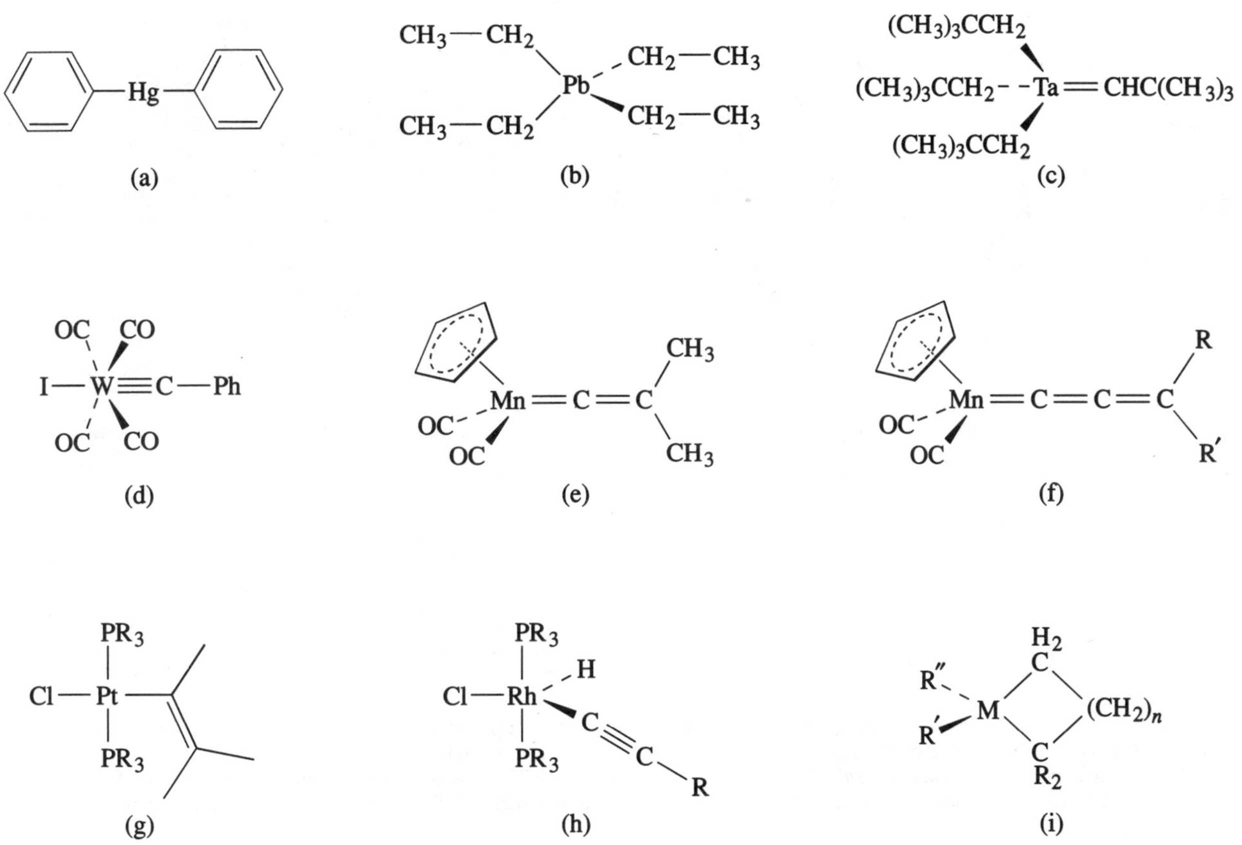

Figure 1 Examples of organometallic compounds containing *monohapto* ligands: (a) diphenylmercury contains two η^1 phenyl ligands; (b) tetraethyllead contains four η^1 ethyl ligands; (c) *tris*(2,2-dimethylpropyl)-2,2-dimethylpropylidenetantalum contains three *monohapto* alkyl ligands and an η^1 carbene (*monohapto* 2,2-dimethylpropylidene) ligand, $=CR_2$; (d) tetracarbonyliodo(phenylmethyne)tungsten contains four *monohapto* η^1 CO ligands and an η^1 carbyne ligand, $\equiv C-R$; (e) a vinylidene complex, containing the η^1 ligand $=C=CR_2$, together with one η^5 cyclopentadienyl ligand (cp) and two η^1 CO ligands; (f) allenylidene, $M=C=C=CRR'$, complexes contain an η^1 carbene ligand; (g) an alkenyl (vinyl) complex, in which the alkene ligand is bound through only one carbon to the metal (η^1); (h) an alkynyl (alkynide) complex, in which the alkyne ligand ($-C\equiv C-R$) is η^1 bound; (i) a metallocycle (the metal is part of a conjugated carbocyclic ring), in which the metal is η^1 bound at the ends of the bidentate organic ligand; note that n is the number of CH_2 units, so that if $n = 0$ the ligand is η^2-bound CH_2CR_2.

☐ Are all the ligands in Figure 1 monohapto? If not, identify the exceptions and state their hapticity.

■ All the ligands in Figure 1 are *monohapto* except the *pentahapto* cyclopentadienyl rings in (e) and (f).

The compounds in Figure 2 provide further illustrations of the systematic names and formulae of *dihapto* and *polyhapto* species.

Figure 3 gives examples of compounds containing bridging ligands. These are identified in the formulae by the prefixed Greek letter μ (mu). The subscript number following μ shows how many metal atoms are bridged by that ligand.

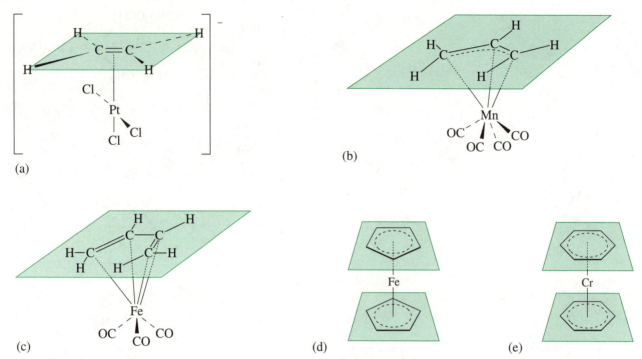

Figure 2 Examples of organometallic compounds containing *polyhapto* ligands: (a) ethenetrichloroplatinate(II) anion, $[PtCl_3(\eta^2\text{-}C_2H_4)]^-$, contains a *dihapto* ethene ligand, $H_2C=CH_2$; (b) allyltetracarbonylmanganese, $[Mn(\eta^3\text{-}C_3H_5)(CO)_4]$, contains a *trihapto* allyl ligand, H_2CCHCH_2; (c) butadienetricarbonyliron contains a *tetrahapto* butadiene ligand, $CH_2=CH-CH=CH_2$; (d) ferrocene (dicyclopentadienyliron), $[Fe(\eta^5\text{-}C_5H_5)_2]$, contains two *pentahapto* ligands, C_5H_5; (e) dibenzenechromium, $[Cr(\eta^6\text{-}C_6H_6)_2]$, contains two *hexahapto* ligands, C_6H_6.

For example, the bridging methyl groups of hexamethyldialuminium, $Al_2(CH_3)_6$, or of polymeric dimethylmagnesium, $\{Mg(CH_3)_2\}_n$, are μ_2-bonded ligands, whereas the methyl groups of the methyl-lithium tetramer†, $(LiCH_3)_4$, are μ_3-bonded because each is directly linked to three metal atoms. Strictly speaking, we should refer to $Al_2(CH_3)_6$ as $bis\{\mu_2\text{-methyldimethylaluminium}(0)\}$.

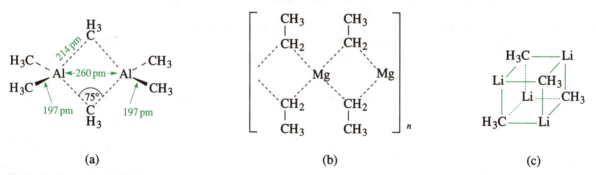

Figure 3 Examples of bridging (μ-bonded) ligands: (a) hexamethyldialuminium, $[Al_2(CH_3)_6]$, contains four terminal (*monohapto*) methyl groups and two bridging (*monohapto*, μ_2-bonded) methyl groups; short-dash broken lines are used to indicate μ-bonded alkyls; (b) diethylmagnesium, $[Mg(CH_2CH_3)_2]$, has a polymeric structure in which all the methylene groups are bridging (*monohapto*, μ_2-bonded); (c) all four methyl groups of the methyl-lithium tetramer, $(LiCH_3)_4$, are *monohapto* and triply bridging (μ_3-bonded).

The three main coordination modes of the carbon monoxide ligand, namely η^1, μ_2 and μ_3, are illustrated in Figure 4. This indicates that a specific ligand can have different coordination modes, dependent on the particular nature of a complex.

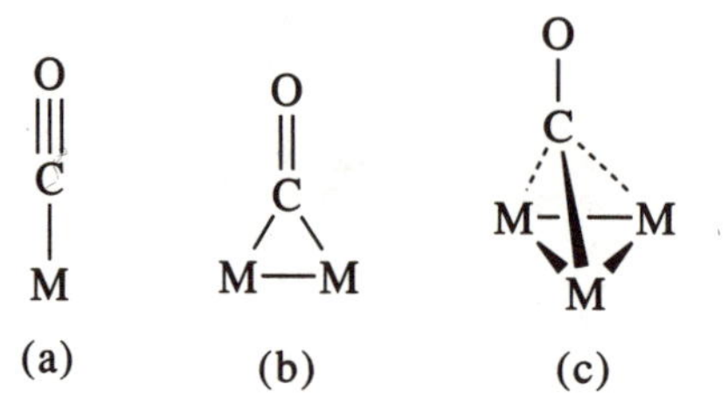

Figure 4 Coordination modes of carbon monoxide: (a) terminal; (b) bridging (μ_2); (c) triply bridging (μ_3).

† Green lines will be used in structures in this Block (such as Figure 3c) whenever the geometry of a complex is being emphasised rather than the bonding.

□ Why is [Hg(η^1-C$_6$H$_5$)$_2$] (Figure 1a) considered to have *monohapto* phenyl ligands and yet [Cr(η^6-C$_6$H$_6$)$_2$] (Figure 2e) is considered to have *polyhapto* η^6 benzene ligands?

■ In [Hg(η^1-C$_6$H$_5$)$_2$] the phenyl rings are bound to the metal by a single carbon atom, so the ligands are *monohapto* or η^1 bound. In [Cr(η^6-C$_6$H$_6$)$_2$] all six carbon atoms of each benzene ring are considered to be bonded to the chromium atom. Hence each benzene ring is a *polyhapto* ligand and is η^6 bound.

By studying the structures and formulae shown in Figures 1–3, you should obtain some feeling for the range of structural types that are known, and the way structural features can be indicated in the formulae. Note that the *monohapto* systems illustrated in Figure 1 include not only alkyl or aryl groups capable of forming single bonds to metal atoms, but also groups capable of forming double or even triple bonds to metal atoms.

Whereas the ligands shown in Figure 1 are monovalent, divalent or trivalent organic groups, normally only encountered when bound to a metal (for example free Ph—C^{3-} is unknown), the ligands shown in Figure 2 include familiar organic molecules such as ethene, buta-1,3-diene and benzene, besides groups such as allyl or cyclopentadienyl which do not have an independent existence.

Organometallic chemistry can trace its history back to 1831 when the Danish pharmacist W. C. Zeise bubbled ethene through a solution of [PtCl$_4$]$^{2-}$, and isolated a compound that was much later identified as K[PtCl$_3$(C$_2$H$_4$)] (Figure 2a). At the turn of the nineteenth century, F. Hein synthesised a new complex by reacting CrCl$_3$ with PhMgBr (a Grignard reagent). This was incorrectly analysed at the time as a phenyl chromium compound, but we now know that the product of the reaction was chromacene, [Cr(η^6-C$_6$H$_6$)$_2$]. In the early 1950s organometallic chemistry blossomed into a new discipline of chemistry, intermediate between the traditional areas of organic and inorganic chemistry. As we shall see later, this growth was stimulated by the analysis of ferrocene (Section 7.1) as a sandwich structure, and, more crucially, by the discovery of β-hydrogen-stabilised ligands (Section 3.1). The availability of new structural tools such as n.m.r. spectroscopy and X-ray crystallography complemented these findings and allowed definitive characterisation of organometallic compounds.

SLC 1

□ Compare the structure of hexamethyldialuminium, Al$_2$(CH$_3$)$_6$, in Figure 3a with those of related molecules of general formula M$_2$X$_6$ with which you are already familiar from a Second Level Course. What types of bond do you think the aluminium alkyl contains?

■ The structure of Al$_2$(CH$_3$)$_6$ resembles those of Al$_2$Cl$_6$ and B$_2$H$_6$. An electron count shows that Al$_2$(CH$_3$)$_6$ contains only six electron pairs to hold its six carbon and two aluminium atoms together, and so resembles Al$_2$Cl$_6$ and B$_2$H$_6$ (which also contain just six bond pairs). Bridging alkyl groups, like bridging hydrogen atoms, are 'electron-deficient' bridges, in that they contain fewer electrons than would be required for formal electron-pair bonds.

The illustrations in Figures 1 to 3 show that individual compounds may contain more than one type of organic ligand, for example both singly and multiply bonded *monohapto* ligands as in the case of the *carbene* complex in Figure 1c, or both bridging and terminal ligands, as in the case of Al$_2$(CH$_3$)$_6$ in Figure 3a. Vast numbers of compounds containing both *monohapto* and *dihapto* or *polyhapto* ligands are also known. Because of this, it is not possible to classify organometallic compounds as belonging exclusively to one structural type or another, though it is useful to focus attention on the properties conferred on a molecule by the presence of a particular type of organic ligand. In our discussions, we shall concentrate on the properties of compounds containing *monohapto* ligands before turning to those containing *dihapto* or *polyhapto* ligands.

2.2 Structural types and the Periodic Table

As metals differ markedly in their capacities to bond to different types of ligand, we should first see how the position of a metal in the Periodic Table provides a guide to the type of organometallic compound it can form. From Figure 5, you will see that the main-Group metals form three categories of organo compound. The most electropositive metals, such as the heavier alkali or alkaline earth metals, and the lanthanides, form essentially ionic alkyls or aryls, typified by sodium or potassium compounds of general formula $M^+ R^-$, in which the organic ligand carries the negative charge. Metals rather less electropositive than these, such as lithium, beryllium, magnesium and aluminium, form organo compounds that are rather more covalent in character. Their alkyls and aryls tend to associate through bridging (μ-bonded) alkyl or aryl ligands, and the same metals also show a limited capacity to form complexes containing *dihapto* or *polyhapto* ligands. The remaining, less electropositive, main-Group metals form organo compounds in which the organic ligands are normally terminally bonded, *monohapto* ligands.

The capacity to form complexes with *dihapto* or *polyhapto* organic ligands is most marked for transition metals, particularly those near the middle of the transition series. Transition metals also have a significant chemistry of complexes containing *monohapto* ligands, and some of them provide further examples of bridge-bonded systems.

SAQ 1 Consider the way in which metal hydrides can be classified according to the position of the metals in the Periodic Table, and compare this with Figure 5. What correlations emerge?

2.3 Bonding in organometallic compounds

From the above discussion, it is apparent that there is a wide range of bond types in organometallic compounds. However, as we have already stated, for a compound to be classified as organometallic the M—C bond must be polarised in the direction $M^{\delta+}-C^{\delta-}$, and this polarisation can vary between two markedly different extremes. At one end, the alkyls and aryls of the most electropositive metals form essentially ionic lattices, in which metal cations and the carbanions are stacked in contact with each other, with relatively little localisation of electronic charge between metal and adjacent carbon atoms. At the other extreme, however, localised electron-pair bonds suffice to describe the bonding in compounds such as tetramethylgermanium, $Ge(CH_3)_4$, the molecular structure of which consists of a tetrahedral array of methyl groups about the central germanium atom; it is comparable with that of 2,2-dimethylpropane, $C(CH_3)_4$. The four metal–carbon bonds of $Ge(CH_3)_4$, though slightly polarised $Ge^{\delta+}-C^{\delta-}$, are single σ covalent bonds of the type familiar throughout organic chemistry.

Between these extremes lies a variety of structural types whose atomic groupings (Figures 2 and 3) indicate some degree of localisation of electronic charge between sets of atoms — that is, some degree of covalency — although localised two-centre electron-pair bonds do not always give an adequate description of the bonding. For example, in the case of the bridging ligands shown in Figure 3, there are not enough electrons to allocate a pair to each metal–carbon link.

The description of the bridge bonding in $Al_2(CH_3)_6$ in terms of two-electron three-centre Al—C—Al bonds allows us to understand why the bridging aluminium–carbon links (214 pm) are longer than the terminal ones (197 pm), since the bridging aluminium–carbon links are multicentre. (The number of electrons that are normally associated with one bond is here associated with two bonds.) It also explains the relatively short (260 pm) distance between the metal atoms, which is compatible with some degree of metal–metal bonding (see Block 4, Section 4.5; 260 pm is only just over twice the aluminium covalent radius). A much longer metal–metal distance (340 pm) is found in the dialuminium hexachloride molecule, Al_2Cl_6. The bridging chlorine atoms in this compound have lone pairs of electrons which form dative bonds; Al_2Cl_6 accordingly adopts a structure in which overlap of metal orbitals does not contribute significantly to the bonding.

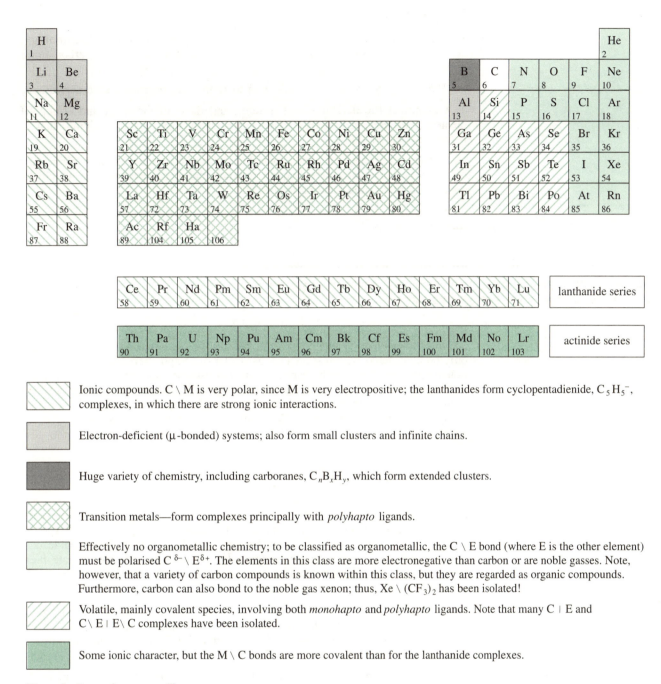

Figure 5 Types of organometallic compound and the Periodic Table.

☐ How many resonances would you expect to see in the ¹H n.m.r. spectrum of $Al_2(CH_3)_6$?

■ Two — one due to the terminal methyls and one due to the bridging methyls. However, this is only the case in solution at or below the coalescence temperature of 205 K. On warming to room temperature, only one resonance is observed owing to rapid methyl exchange.

The structure of the methyl-lithium tetramer in Figure 3c is an indication of how the metal atoms try to use as many orbitals as possible. There is a **four-centre two-electron bond** by which each methyl group bridges three lithium atoms.

Not only can organometallic compounds contain localised (C—Ge) and ionic bonding (C⁻ K⁺), but also multiply bound carbons as in the carbene (=CR₂) and carbyne (≡C—R) ligands, where π-bonding is important. Furthermore, in Section 5 we shall see how reactive C=E (E = Si, Se, Te, P, As, Bi) multiple bonds can be made and stabilised by coordination to a transition metal.

2.4 Summary of Section 2

1 Organometallic compounds can be classified by the *hapto* terminology, according to which the *hapticity* of a ligand refers to the number of carbon atoms bound to a metal.

2 The most electropositive metals tend to form ionic organometallic derivatives.

3 Organometallic derivatives of lithium, and those of Group II and Group III elements, are generally more covalent. They frequently contain bridging (that is, μ-bonded) carbon atoms.

4 The idea that chemical bonds consist of pairs of electrons shared between pairs of atoms is inadequate for μ-bonded organometallic compounds, for which the concept of multicentre bonding is needed.

5 Complexes with *dihapto* and *multihapto* ligands are mainly formed by transition metals.

3 STABILITY AND REACTIVITY OF *MONOHAPTO* ORGANOMETALLIC DERIVATIVES

As with other types of compound, the stability of organometallic compounds can only be considered with respect to a particular decomposition route. Thus, a compound may be stable with respect to its constituent elements, but not to other decomposition products or to specified reagents. Stability on heating, reaction with water and with dioxygen are significant factors to consider with organometallic compounds, and both thermodynamic and kinetic aspects need to be addressed.

3.1 Thermal stability

The thermodynamic stability of a compound relative to its constituent elements (or relative to the products of specific reactions) can be deduced if we know the standard Gibbs function, or free energy, of formation, $\Delta G_f^\ominus$, of all the species concerned. Even if we do not know the appropriate $\Delta G_f^\ominus$ values (few reliable data are available for organometallic compounds), we may be able to estimate them from known standard enthalpies of formation $\Delta H_f^\ominus$ and by estimating the likely contributions from the entropy-related term $T\Delta S^\ominus$.

From comparison of thermodynamic data, decomposition of metal methyl groups into metal, carbon and dihydrogen would be an exothermic process for such heavy-metal alkyl species as $Bi(CH_3)_3$, $Pb(CH_3)_4$ and $Hg(CH_3)_2$, but endothermic for semi-metal alkyls such as $Si(CH_3)_4$, $B(CH_3)_3$ and $Ge(CH_3)_4$.

In practice, the thermal decomposition of metal alkyl species commonly gives the metal, dihydrogen and mixed hydrocarbons, rather than elemental carbon. Hence if we wish to determine whether these reactions are endothermic or exothermic, we have to take account of the enthalpies of formation of the hydrocarbons formed.

☐ For example, write a balanced equation for the decomposition of tetramethyllead to give lead metal, methane and ethene.

■ The equation is:

$$Pb(CH_3)_4(g) = Pb(s) + 2CH_4(g) + C_2H_4(g) \qquad 1$$

A detailed analysis of the thermodynamic data shows that although $\Delta H_m^\ominus$ does depend markedly on the particular hydrocarbons formed (the reaction would be much less exothermic if ethene were the only hydrocarbon product), we nevertheless conclude that tetramethyllead is unstable with respect to metal and mixed hydrocarbons. Analysed in this way, organometallic compounds like $B(CH_3)_3$ and $Si(CH_3)_4$ are stable to such decomposition, whereas the compounds $Cd(CH_3)_2$, $In(CH_3)_3$ and $Bi(CH_3)_3$, among others, resemble $Pb(CH_3)_4$ in being unstable. Nevertheless, all of these compounds are obtainable and can be stored without decomposition at room temperature.

☐ Why do you think this is?

■ This is because even though the compounds are thermodynamically unstable, they have no route with low-enough activation energy by which to decompose at room temperature. Their decomposition is thus *kinetically controlled*.

This point is illustrated schematically in Figure 6, which represents the decomposition of a metal alkyl, MR_n, by homolytic dissociation of its metal–carbon bonds. Although the decomposition of a metal alkyl into metal, dihydrogen and various hydrocarbons may be thermodynamically favourable ($\Delta G_m^\ominus$ in Figure 6 is negative), initiation of the reaction requires energy ($G^\ddagger$ in Figure 6) for the first stage of the reaction, which is the formation of radicals $MR_{n-1}\cdot$ and $R\cdot$. Radicals are highly reactive species with an unpaired electron; in order to generate radicals, energy must be supplied to achieve homolytic fission of the M—C bond into the M· and C· fragments.

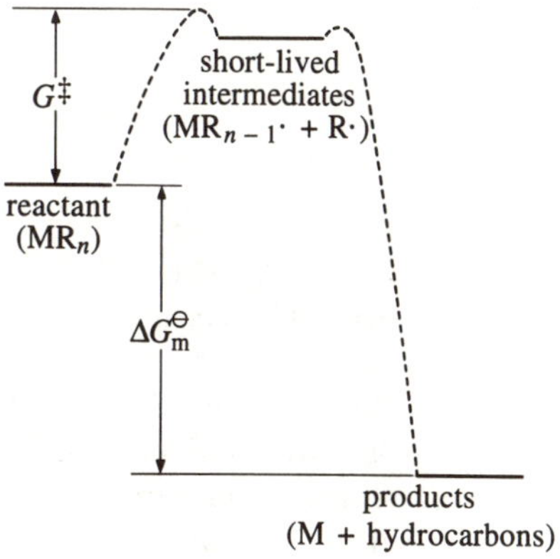

Figure 6 Schematic representation of the thermal decomposition of a metal alkyl, [MR_n], by unimolecular homolytic dissociation. ($\Delta G_m^\ominus$ represents the standard Gibbs free energy change for the overall decomposition, and so indicates the thermodynamic (in)stability of MR_n. $G^\ddagger$, the activation energy of the first stage, is a measure of the kinetic stability of MR_n.)

☐ What will be the major factor in determining $G^\ddagger$?

■ The size of $G^\ddagger$ will depend largely on the metal–carbon bond strength — that is, on the value of $E_m(M-C)$.

For the Group IV methyls, as the metal–carbon bond strengths decrease down the Group (Figure 7), thermal decomposition becomes progressively easier on both kinetic and thermodynamic grounds.

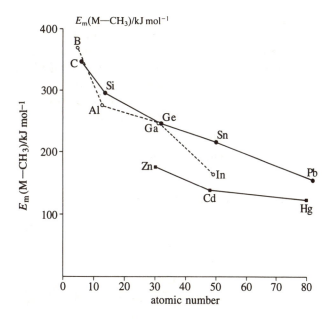

Figure 7 Group trends in average molar bond enthalpies, $E_m(M-C)$, for metal methyls, [$M(CH_3)_n$].

In homolytic dissociation reactions of this sort, once the radicals $MR_{n-1}\cdot$ and $R\cdot$ are formed, they will be very reactive and so will rapidly react further to generate stable products, for example by *combination* to form R_2, or by hydrogen elimination or abstraction reactions to form an alkane, RH, and an alkene (R minus H) — a *disproportionation* reaction. It is the thermodynamic stability of these products with respect to the radicals that provides the driving force for the decomposition. This is illustrated schematically for Si(CH$_2$CH$_3$)$_4$ in Figure 8.

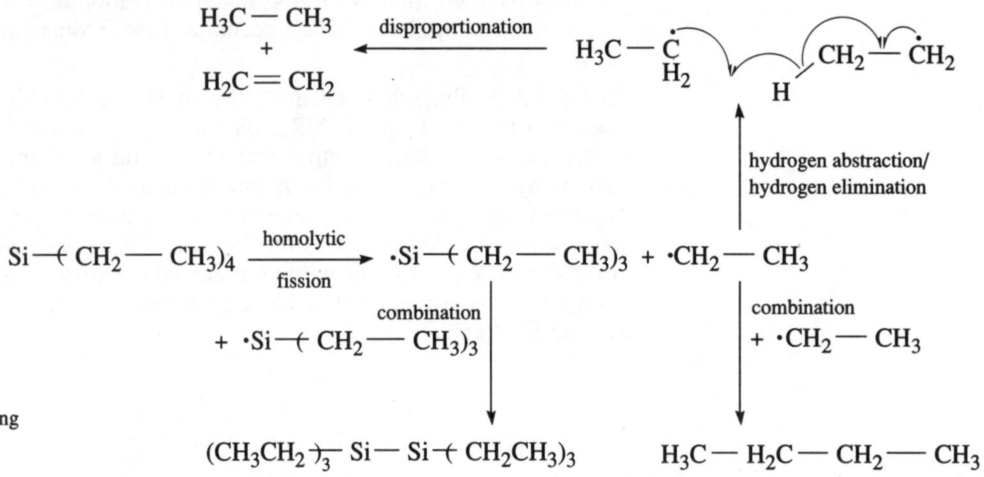

Figure 8 Radical reactions resulting from homolytic dissociation of [Si(CH$_2$CH$_3$)$_4$].

Homolytic dissociation of metal–carbon bonds is one of several possible ways in which organometallic compounds can decompose thermally. Other decomposition reactions may involve carbon–hydrogen bonds. One particularly important source of thermal instability is the presence of C—H groups separated by one carbon atom from the metal — that is, β-hydrogen atoms — as in compounds containing the structural unit MCR^{1}R^{2}CHR^{3}R^4 (where R^1, R^2, R^3 and R^4 may be other organic groups or hydrogen atoms). Such compounds can decompose by migration of the β-hydrogen atom to the metal, leading to formation of a metal hydride and the elimination of an alkene:

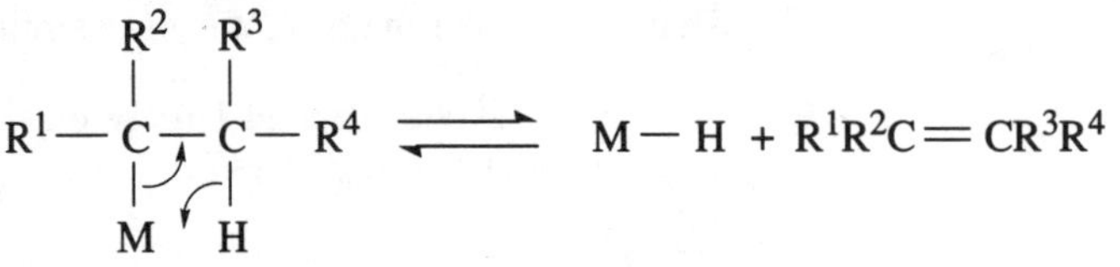

$$R^1-\underset{\underset{M}{|}}{\overset{\overset{R^2}{|}}{C}}-\underset{\underset{H}{|}}{\overset{\overset{R^3}{|}}{C}}-R^4 \rightleftharpoons M-H + R^1R^2C=CR^3R^4 \qquad 2$$

β-elimination reactions proceed more readily if the metal atom has a tendency to increase its coordination number. Once again, this tendency increases down a Group in the Periodic Table.

Two other degradation reactions that are related to β-elimination should be mentioned at this stage: these are **orthometallation** and **α-hydrogen elimination**. In the former process a proton is transferred to the metal from an *ortho* position of an aromatic ring (Figure 9a); for α-hydrogen elimination the hydrogen is transferred to the metal, which is normally in a high oxidation state (Figure 9b). Orthometallation occurs when a C—H bond is located close to the metal atom. Both reactions require a metal atom that has room to coordinate further ligands.

Both orthometallation and α-hydrogen elimination are important in the catalytic activity of some organometallic processes.

3.2 Stability to oxidation

Organometallic compounds are thermodynamically unstable to oxidation as a result of the large negative standard free energies of formation of metal oxide, carbon dioxide and water. In fact, the combustion of organometallic compounds is a highly exothermic process, for example

$$Zn(CH_3)_2(g) + 4O_2(g) = ZnO(s) + 2CO_2(g) + 3H_2O(g);$$
$$\Delta H_m^\ominus = -1\,918 \text{ kJ mol}^{-1} \qquad 3$$

$$Sn(CH_3)_4(g) + 8O_2 = SnO_2(s) + 4CO_2(g) + 6H_2O(g);$$
$$\Delta H_m^\ominus = -3\,591 \text{ kJ mol}^{-1} \qquad 4$$

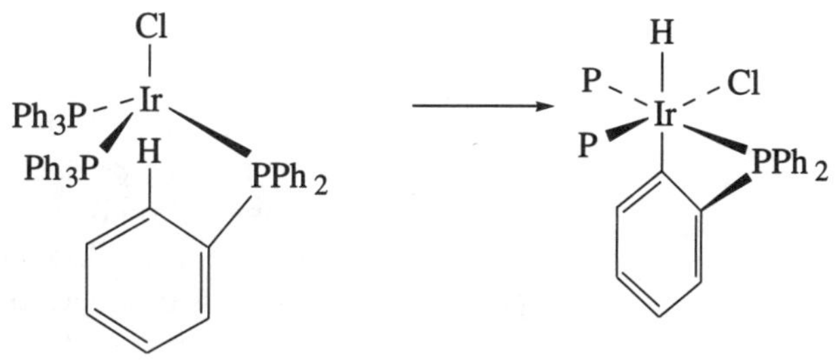

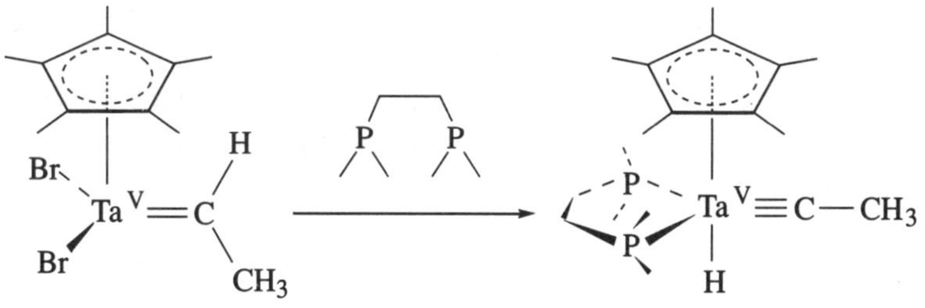

Figure 9 (a) Orthometallation: the proton *ortho* to the position on an aromatic ring which is bonded to a metal is transferred to the metal; in this example there is simultaneous formation of an Ir—C bond. (b) α-Hydrogen elimination: the hydrogen of a carbene is transferred to the metal.

Despite the exothermic character of their oxidation reactions, many organometallic compounds, such as the alkyls and aryls of mercury, silicon, germanium, tin and lead, are kinetically inert to air or oxygen at room temperature; others are kinetically unstable to aerial oxidation. The latter compounds either fume immediately, or even burst into flames on exposure to air. This type of behaviour is exhibited by the lighter alkyls of Groups I and II, and all the alkyls of Group III: Li, Na, Be, Mg, Zn, B, Al, Ga, In and Tl.

3.3 Stability to hydrolysis

The hydrolysis of organometallic compounds commonly involves nucleophilic attack by water, and (like oxidation) is helped by the presence of empty low-energy orbitals on the metal. Thus, the organo derivatives of elements of Groups I and II, plus Zn, Cd, Al, Ga and In, are readily hydrolysed; the rates of hydrolysis depend on the polarity of their metal–carbon bonds. Highly polar bonds $M^{\delta+}-C^{\delta-}$, as in $Al_2(CH_3)_6$ or $(LiCH_3)_4$, are so rapidly cleaved by water that an explosively violent reaction may occur. By contrast, trimethylborane, $B(CH_3)_3$, whose boron–carbon bonds have only slight polarity, is unaffected by water at room temperature even though, in principle, the empty 2p orbital on the boron atom affords a site for nucleophilic attack. Neutral organo derivatives of transition metals with similarly filled shells of electrons are also inert to hydrolysis.

SAQ 2 Tetrachlorosilane, $SiCl_4$, hydrolyses rapidly on exposure to moisture, but tetramethylsilane, $Si(CH_3)_4$, is unaffected. Why is the behaviour of these two compounds so different?

3.4 General reactivity

☐ The following list contains factors that can affect reactivity. Examine each factor and decide whether it is a contribution to the thermodynamic stability of a compound or to its kinetic stability (or **lability** as is more often applied to kinetic reactivity). Factors may be related to either or both circumstances.

- enthalpy of formation of the products
- polarity of bonds
- availability of empty low-energy orbitals
- lone pairs of electrons
- metal–carbon bond energies
- metal's ability to coordinate further ligands

■ The factors that determine whether a reaction is thermodynamically favourable relate to the difference in energy between the reactants and the products. The enthalpy of formation of the products is clearly important; if the enthalpy of formation of the products is large and negative, then the reaction is likely to be favourable. Similarly, if the metal–carbon bond energy is very high, then that can have a stabilisation effect. The M—C bond energy is also implicated in the kinetics of a reaction, as described below. The other factors on the list all relate to the rate at which a reaction will take place. They are factors affecting the energy of the transition state. To a large extent, the M—C bond energy in a homolytic reaction determines the rate at which the reaction will go. The extent of bond polarity determines how easy it is for a nucleophile to attack at $M^{\delta+}$, or an electrophile to attack at $C^{\delta-}$. The availability of empty low-energy orbitals, and the ability to increase coordination both affect the rate at which the M—C bond can be attacked. Lone pairs of electrons are nucleophilic in character and their presence may also affect reactivity.

We can now see why certain Group IV organometallic compounds are among the least reactive organometallic compounds known, and understand the progressively increasing reactivity that is found from silicon to lead in Group IV, or within a Period from a Group IV element to the more electropositive metals in earlier Groups (for example, from Si to Al to Mg to Na), or to the neighbouring Group V element. The reactivities of representative examples of these various Groups are discussed in Section 5. First, however, let us turn to the problem of how to synthesise organometallic compounds.

3.5 Summary of Section 3

1 Many organometallic compounds are thermodynamically unstable with respect to decomposition into metal, hydrocarbons and dihydrogen.

2 The existence of organometallic compounds for which the decomposition reactions are exothermic can be attributed to a high activation energy for the decomposition reaction. They are therefore thermodynamically unstable, but kinetically stable.

3 A common decomposition route for organometallic compounds is β-hydrogen migration (reaction 2).

4 Some organometallic compounds are kinetically unstable to aerial oxidation.

5 Kinetic instability to hydrolysis is also associated with low-energy empty orbitals, and/or highly polar $M^{\delta+}$—$C^{\delta-}$ bonds.

6 Chemical reactivity of organometallic compounds is associated with: (a) weak or highly polar metal–carbon bonds; (b) a metal's ability to coordinate further ligands; (c) labile organic groups, for example those with a β-hydrogen.

7 Organometallic compounds can also decompose by α-hydrogen elimination and orthometallation reactions.

4 METHODS OF FORMING METAL–CARBON BONDS

There are four main methods for producing organometallic compounds with *monohapto* ligands.

(a) Reactions of a metal with an alkyl or aryl haloalkane (**metallation**):

$$2M + nRX \longrightarrow MR_n + MX_n \qquad 5$$

for example

$$2Li + BuCl \longrightarrow LiCl + BuLi \qquad 6$$

(b) Reaction of a metal with an organo derivative of another metal (**transmetallation**):

$$M' + MR \longrightarrow M'R + M \qquad 7$$

for example

$$2Ga + 3Hg(CH_3)_2 \longrightarrow 2Ga(CH_3)_3 + 3Hg \qquad 8$$

(c) Reaction of a metal halide with the organo derivative of another metal (an example of **metathesis**, which means an exchange of ligands between groups):

$$MR + M'X \longrightarrow M'R + MX \qquad 9$$

for example

$$3LiCH_3 + TlCl_3 \longrightarrow Tl(CH_3)_3 + 3LiCl \qquad 10$$

(d) Insertion of alkenes or alkynes into metal–hydrogen bonds (**hydrometallation**):

$$M-H + R^1R^2C=CH_2 \longrightarrow M-CH_2-CR^1R^2H \qquad 11$$

For example

$$\text{'BH}_3\text{'} + 3CH_3CH=CH_2 \longrightarrow B(CH_2CH_2CH_3)_3 \qquad 12$$

4.1 Reaction of a metal with a haloalkane — metallation

Metallation is a suitable method for the laboratory synthesis of organo derivatives of the more reactive metals, such as lithium, magnesium (where it provides the standard route to Grignard reagents) and aluminium. For the preparation of organo derivatives of less-reactive metals or semi-metals (such as As or Bi), it can be modified by the use of an alloy instead of the pure metal; for example, the petrol antiknock compound tetraethyllead is made by reaction of a sodium–lead alloy with chloroethane:

$$4CH_3CH_2Cl + 4Na/Pb \longrightarrow Pb(CH_3CH_2)_4 + 4NaCl + 3Pb \qquad 13$$

The thermodynamic driving force for the reaction (which involves the cleavage of a carbon–halogen bond and the formation of an organometallic compound, which may be of low thermodynamic stability relative to the metal; see Table 1) is mainly provided by the large negative enthalpy of formation of the metal halide.

Table 1 Standard molar enthalpies of formation of some metal chlorides* and the differences† between these and the standard molar enthalpies of formation of the corresponding metal methyls

M	Group II; MCl$_2$ $\Delta H_f^\ominus$	diff.	M	Group III; MCl$_3$ $\Delta H_f^\ominus$	diff.	M	Group IV; MCl$_4$ $\Delta H_f^\ominus$	diff.	M	Group V; MCl$_3$ $\Delta H_f^\ominus$	diff.
			B	−403‡	−281	C	−135§	+32			
			Al	−705	−617	Si	−687§	−449	P	−320§	−224
Zn	−416	−471	Ga	−525	−486	Ge	−690§	−582	As	−305§	−320
Cd	−391	−501	In	−537	−709	Sn	−511§	−492	Sb	−382	−413
Hg	−230	−323	Tl	—	—	Pb	−314‡	−451	Bi	−379	−571

*$\Delta H_f^\ominus = \Delta H_f^\ominus (MCl_n) / \text{kJ mol}^{-1}$.

† diff. $= \Delta H_f^\ominus (MCl_n) / \text{kJ mol}^{-1} - \Delta H_f^\ominus (MMe_n) / \text{kJ mol}^{-1}$.

‡ Relates to gaseous MCl_n.

§ Relates to liquid MCl_n; all other values relate to solid MCl_n.

4.2 Reaction of a metal with an organo derivative of another metal — transmetallation

Endothermic or weakly exothermic organometallic compounds, $M(CH_3)_n$, are expected to prove the most versatile reagents in transmetallation reactions. Examples of such reagents are the alkyl derivatives of the heavy main-Group metals Hg, Tl, Pb and Bi. Of these, the air-stable and readily prepared alkylmercury compounds HgR_2 have been widely used, even though they are toxic.

Metals that can be alkylated (or arylated) by organomercury compounds include the alkali metals, alkaline earth metals, plus Zn, Al, Ga, Sn, Pb, Sb, Bi, Se and Te. In some cases, formation of a metal amalgam assists the reaction. Note that, although the more electropositive metal normally displaces the less electropositive metal from its organo substituents in such reactions, the controlling factor is the standard Gibbs free-energy change in the reaction.

4.3 Reaction of a metal halide with an organo derivative of another metal — metathetical reactions

The third main method of preparing organometallic derivatives is *metathesis*, in which the partners X and R are exchanged between the metals:

$$MX + M'R \longrightarrow MR + M'X \qquad 9$$

The direction in which the reaction is likely to go can be inferred by considering the enthalpy changes involved because the entropy changes would be expected to be insignificant.

The standard molar enthalpies of formation of some main-Group metal chlorides are given in Table 1, together with the differences between these and the standard molar enthalpies of formation of the methyl derivatives. Within any one Group, a high negative value in the difference column implies that the methyl derivative of that element is likely to be a more powerful methylating agent than the methyl derivative of an element with a low negative value in the difference column; that is, the former should be capable of methylating the chloride of the latter. For example, consider reaction 14:

$$Hg(CH_3)_2(g) + ZnCl_2(s) = Zn(CH_3)_2(g) + HgCl_2(s) \qquad 14$$

Using the data in Table 1, the enthalpy change, $\Delta H_m^\ominus$, for the reaction in the direction shown is $-323 + 471 = +148\,\text{kJ mol}^{-1}$; that is, the reaction is endothermic. Dimethylmercury cannot therefore be used to methylate zinc chloride.

Instead, dimethylzinc could be used to prepare dimethylmercury from $HgCl_2$. Similarly in Group III, $Al(CH_3)_3$ can be used for the preparation of $B(CH_3)_3$ from BCl_3.

☐ From the data in Table 1, deduce which of the metal methyls is thermodynamically the most powerful methylating agent.

■ This can be calculated by expressing the difference, $\Delta H^\ominus_f(MCl_n) - \Delta H^\ominus_f(M(CH_3)_n)$, in kJ per mole of methyl groups; that is

$$\frac{1}{n}[\Delta H^\ominus_f(MCl_n) - \Delta H^\ominus_f(M(CH_3)_n)]$$

Dimethylcadmium (difference −251 kJ per mole of methyl groups) is the most powerful, and $Zn(CH_3)_2$ and $In(CH_3)_3$ come next, both with differences of −236 kJ per mole of methyl groups.

☐ From the data in Table 1, deduce whether trimethylborane is a suitable reagent for the preparation of the Group IV tetramethyls, $M(CH_3)_4$, from the tetrachlorides, MCl_4 (M = Si, Ge, Sn, Pb).

■ The standard enthalpy changes for the reactions

$$4B(CH_3)_3 + 3MCl_4 = 3M(CH_3)_4 + 4BCl_3 \qquad 15$$

are positive for M = Si, Ge, Sn or Pb, showing that the reverse reaction is thermodynamically preferred. For example, for M = Si, the standard enthalpy change for the above reaction is $-(4 \times 281) + (3 \times 449) = +223$ kJ mol^{-1}. Tin alkyls and aryls, SnR_4, have in fact been used for the preparation of organoboranes, BR_3, from BCl_3.

The most versatile alkylating or arylating reagents are the organo derivatives of the more electropositive metals. As lithium and magnesium reagents, for example butyllithium and phenylmagnesium bromide (an example of a Grignard reagent), are readily available or easily preparable, these reagents are the ones most commonly used in the laboratory; aluminium alkyls tend to be preferred for industrial-scale work, since by using them the costly ether solvents necessary for Grignard syntheses can be avoided.

Organolithium reagents are more reactive than Grignard reagents (general formula RMgX, R = alkyl or aryl, X = halogen), because they are less bulky and more nucleophilic. Hence they are preferred in cases where complete substitution of halogen by alkyl groups is required. For example, whereas Grignard reagents replace only two of the three chlorine atoms of $TlCl_3$, lithium reagents replace all three:

$$R_2TlCl \xleftarrow{RMgX} TlCl_3 \xrightarrow{LiR} TlR_3 \qquad 16$$

4.4 Insertion of alkenes or alkynes into metal–hydrogen bonds — hydrometallation

The fourth main route to metal alkyls, the insertion of alkenes into metal–hydrogen bonds, is important in organotransition-metal chemistry and also for the synthesis of organoboron, organoaluminium and organosilicon compounds (Section 5):

$$M-H + R^1R^2C{=}CH_2 \longrightarrow M-CH_2-CR^1R^2H \qquad 11$$

Other insertion reactions of alkenes or alkynes, into metal–nitrogen, metal–phosphorus, metal–oxygen, metal–halogen or metal–metal bonds, also find occasional use as routes to metal–carbon-bonded species; for example

$$\text{Ph}_3\text{Sn}-\text{PPh}_2 + \underset{\text{Ph}\ \ \ \text{H}}{\overset{\text{Ph}\ \ \ \text{H}}{\text{C}=\text{C}}} \longrightarrow \text{Ph}_3\text{Sn}-\text{CH}_2-\underset{\text{Ph}}{\overset{\text{H}}{\text{C}}}-\text{PPh}_2 \qquad 17$$

Insertions into other metal–carbon bonds form the basis of the catalytic activity of organometallic compounds towards alkene polymerisation; for example

$$\text{AlR}_3 + n\text{C}_2\text{H}_4 \longrightarrow \text{Al}\begin{cases}(\text{C}_2\text{H}_4)_x\text{R}\\(\text{C}_2\text{H}_4)_y\text{R}\\(\text{C}_2\text{H}_4)_z\text{R}\end{cases} \qquad 18$$

$(n = x + y + z)$

High pressures (about 10 MPa) are used in order to favour this reaction over the alternative reactions of alkene elimination and hydride formation.

SAQ 3 Suggest reagents that would be suitable for the laboratory synthesis of the following compounds: (a) $SiMe_4$; (b) $EtMgCl$; (c) BEt_3; (d) $TlPh_2Cl$; (e) $SnEt_4$.

4.5 Summary of Section 4

1 The main methods of preparing organometallic compounds containing *monohapto* ligands are metallation (M + RX), transmetallation (M + M'R), metathesis (MR + M'X), and hydrometallation (MH + alkene or alkyne).

2 Transmetallation depends for its success on the difference between the standard Gibbs free energy of formation of MR and M'R.

3 The most widely used route to *monohapto* derivatives is the reaction between a metal halide and an organo derivative of another metal. Table 1 indicates the feasibility of this reaction in certain cases.

4 The hydrometallation of alkenes or alkynes is important in organo transition-metal chemistry and also provides a preparative route to organoboron, organoaluminium and organosilicon compounds.

5 SYSTEMS CONTAINING *MONOHAPTO* LIGANDS

Having touched on general features of organometallic compounds, their structures, bonding, reactions and methods of synthesis, we are now in a position to look more closely at some specific systems, which may be regarded as representative of the various categories of organometallic compound. In Section 2.2 and Figure 5, we grouped compounds containing *monohapto* ligands into certain categories.

☐ What are those categories, and which types of element typically fit each category?

■ In the first category are compounds in which μ-bonding is common, as in the case of derivatives of Li, Be, Mg and Al.

The second category embraces derivatives of the less-electropositive main-Group metals and semi-metals, most of which are volatile, covalent species.

The third category includes *monohapto* organo derivatives of transition metals, where thermal instability and reactivity are common features of compounds with incompletely filled shells of electrons. Examples from these categories are discussed in the following Sections.

5.1 Organomagnesium compounds

Organomagnesium compounds are versatile and are widely used for the conversion of organic functional groups. The most common organomagnesium compounds are Grignard reagents, the general name given to organomagnesium halides, RMgX. The first examples were prepared from Mg and RX by P. A. Barbier and F. A. V. Grignard about the turn of the century, and the value of these compounds as alkylating, arylating and/or reducing agents was demonstrated in subsequent work by Grignard. The significance of his contribution to synthetic organic chemistry was recognised by the award of the Nobel Prize for Chemistry in 1912.

Bromoalkanes are the haloalkanes most often used in the laboratory for preparing Grignard reagents. This is because their reactions require lower activation energies than chloroalkanes, even though the latter appear thermodynamically preferable.

SLC 3 (The synthesis and application of Grignard reagents was discussed in a Second Level Course.) The general equation for the preparation of a Grignard reagent is

$$RX + Mg \longrightarrow RMgX \qquad \qquad 19$$

Despite being less reactive, chloroalkanes are cheaper and are preferred industrially. Fluoroalkanes, RF, are less reactive still, and are unsuitable for Grignard syntheses. Iodoalkanes, other than iodomethane, are rarely used because of their higher cost and because of a tendency for coupling reactions to predominate. In these reactions the alkyl groups combine together to give R_2 rather than RMgI:

$$2RI + Mg \longrightarrow MgI_2 + R-R \qquad \qquad 20$$

Solutions of Grignard reagents in ether solvents can be stored under a dry, oxygen-free atmosphere (for example, under dry N_2). They contain predominantly solvated species, which can be isolated as dietherates, $RMgX.2Et_2O$, on evaporation of solvent. Dimeric species, $(RMgX.Et_2O)_2$, containing bridging halogen atoms, are also present in concentrated solutions, together with some dialkyl, MgR_2, and dihalide, MgX_2, in dynamic equilibrium with RMgX:

$$2RMgX \rightleftharpoons MgR_2 + MgX_2 \qquad \qquad 21$$

The position of this equilibrium varies with R, X, the solvent and the temperature. In diethyl ether, at the concentration (about $0.5-1.0\,\text{mol}\,l^{-1}$ in RMgX) commonly used in Grignard syntheses, the following equilibria are important:

$$\begin{array}{c}\text{R}\quad\text{X}\quad\text{OEt}_2\\ \diagdown\;\diagup\;\diagdown\;\diagup\\ \text{Mg}\quad\text{Mg}\\ \diagup\;\diagdown\;\diagup\;\diagdown\\ \text{Et}_2\text{O}\quad\text{X}\quad\text{R}\end{array} + 2\text{Et}_2\text{O} \rightleftharpoons 2\left[\begin{array}{c}\text{R}\quad\text{X}\\ \diagdown\;\diagup\\ \text{Mg}\\ \diagup\;\diagdown\\ \text{Et}_2\text{O}\quad\text{OEt}_2\end{array}\right] \rightleftharpoons \begin{array}{c}\text{R}\quad\text{R}\\ \diagdown\;\diagup\\ \text{Mg}\\ \diagup\;\diagdown\\ \text{Et}_2\text{O}\quad\text{OEt}_2\end{array} + \begin{array}{c}\text{X}\quad\text{X}\\ \diagdown\;\diagup\\ \text{Mg}\\ \diagup\;\diagdown\\ \text{Et}_2\text{O}\quad\text{OEt}_2\end{array} \qquad 22$$

The greatest use of Grignard reagents in organic synthesis utilises their ability to engage in carbon–carbon bond-forming reactions with compounds containing unsaturated functional groups (C=O, C=N, C≡N, C=S, C=C, C≡C). The products of reactions between Grignard reagents and other compounds can be understood in terms of nucleophilic attack by the metal-attached carbon atom of the Grignard reagent at the most positive centre in the other compound (which we call the **substrate**). The orientation of insertion of the unsaturated groups into the metal–carbon bond can be explained in these terms. For example, the reaction between a Grignard agent and a ketone can be represented as follows:

$$R-MgX + \underset{R''}{\overset{R'}{C}}=O \longrightarrow R-\underset{R''}{\overset{R'}{C}}OMgX \xrightarrow{H_2O} RR'R''C(OH) + Mg(OH)X \qquad 23$$

It also explains the cleavage of the metal–carbon bonds of Grignard reagents, which is effected by protic acids, $H^{\delta+}-X^{\delta-}$ (HOH, HOR, HSH, HCl, HBr, HC≡CR, etc.); for example

$$\left[\begin{array}{c} R^{\delta-}-Mg^{\delta+}X \\ H^{\delta+}-Cl^{\delta-} \end{array}\right] \longrightarrow R-H + MgClX \qquad 24$$

It should be noted that the final stage of any synthesis involving Grignard reagents normally involves protic acid addition. This aspect of the behaviour of Grignard reagents shows that they can be regarded as bases.

Although the polarity of the C—Mg bond explains the orientation of alignment of the reactants, the first step in many cases involves nucleophilic attack by the reagent on the metal atom. Thus, ketones form adducts R′R″CO.RMgX, which subsequently rearrange to alkoxides, RR′R″COMgX. Acid work-up then yields a tertiary alcohol:

$$25$$

Products of radical reactions may also be observed from Grignard reagents, particularly when the magnesium from which it was prepared contains transition-metal impurities. The mechanism envisaged involves electron transfer from a metal–carbon bonding orbital of the reagent into a π^* orbital of the multiply bonded compound, the effect of which may be represented as follows:

$$\underset{R''}{\overset{R'}{C}}=O + \underset{R}{MgX} \longrightarrow \left[\begin{array}{c} R'R''\dot{C}-O \\ \diagdown \\ MgX \\ \diagup \\ R\cdot \end{array}\right] \xrightarrow{-R\cdot} \tfrac{1}{2}\left[\begin{array}{c} R'R''C-OMgX \\ | \\ R'R''C-OMgX \end{array}\right] \qquad 26$$

Grignard reagents have proved to be highly useful synthetic reagents, both in the selectivity and the diversity of the reactions that they can perform (Figure 10).

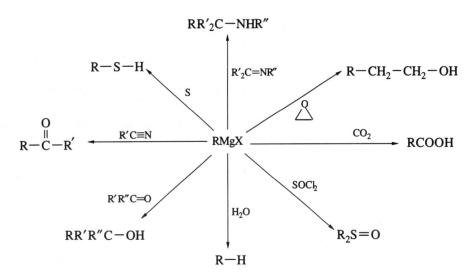

Figure 10 Applications of organomagnesium reagents. (Note that hydrolysis may be required to produce the product shown.)

5.2 Organolithium compounds

Like Grignard reagents, organolithium compounds are reagents commonly used in synthetic organic chemistry, though rarely isolated from the solutions in which they may be prepared (or bought). They can be prepared from haloalkanes and the metal, which is conveniently available in a finely divided form (with a large surface area), set in paraffin oil; for example

$$\text{BuCl} + 2\text{Li} \xrightarrow[\text{petrol solution}]{\text{benzene or}} \text{LiBu} + \text{LiCl} \qquad 27$$

Unlike Grignard reagents, which are very difficult to prepare in non-donor solvents, lithium alkyls can be prepared in hydrocarbon solutions: indeed the use of ether solvents is ruled out for many lithium alkyls, which react with ethers by attack at a β-hydrogen to form lithium ethoxide and RH with elimination of ethene:

$$\text{LiR} + \text{Et}_2\text{O} \longrightarrow \text{RH} + \text{C}_2\text{H}_4 + \text{LiOEt} \qquad 28$$

—CH—CH$_2$—CH$_3$
 |
 CH$_3$
iso-butyl (Bu^i)

CH$_3$
 |
—CH
 |
 CH$_3$
iso-propyl (Pr^i)

CH$_3$
 |
—C—CH$_3$
 |
 CH$_3$
tertiary butyl (Bu^t)

—CH$_2$—CH—CH$_3$
 |
 CH$_3$
secondary butyl (Bu^s)

The readiness with which this reaction occurs increases in the sequence R = Me < Ph < Bu < Et < Pr < Bu^i < Pr^i ≈ Bu^s < Bu^t, so although methyl-lithium can be used as an ether solution (it is virtually insoluble in hydrocarbon solvents), butyllithium (itself a liquid) is normally available as a pentane or hexane solution, in which it is present as tetramers (LiBu)$_4$ or hexamers (LiBu)$_6$ (see Figure 3c for the structure of (LiMe)$_4$).

Lithium alkyls are highly reactive substances, which must be stored and handled under an inert atmosphere (nitrogen or argon, not carbon dioxide, with which they react to form carboxylates, RCO$_2$Li), with rigorous exclusion of moisture. Like Grignard reagents, they are bases and react with substances containing hydrogen attached to, or close to, an activating functional group, to form hydrocarbons:

$$\text{LiR} + \text{HX} \longrightarrow \text{LiX} + \text{RH} \qquad 29$$

where X = OH, OR, NHR, halogen, or an organic group more electronegative than R. Lithium alkyls also add across multiple bonds (C=O, C=N, C=S, C≡N, C≡C) with an orientation that reflects their polarity:

$$\overset{\delta+}{\text{Li}}-\overset{\delta-}{\text{R}} + \overset{\delta+}{\text{R}'}\overset{\delta-}{\text{R}''\text{C}=\text{O}} \longrightarrow \text{R}'\text{R}''\text{RCOLi} \xrightarrow{\text{H}_2\text{O}} \text{R}'\text{R}''\text{RCOH} \qquad 30$$

Organolithium compounds are useful alternatives to Grignard reagents in preparative organic chemistry, having more-polar metal–carbon bonds (lithium is less electronegative than magnesium); they are more reactive than Grignard reagents. The difference in reactivity is particularly marked in cases where the size of substituent groups may restrict the approach of bulky Grignard reagents.

Organolithium compounds, particularly LiBu, are convenient hydrocarbon-soluble bases for use in dehydrohalogenations, as in the synthesis of Wittig reagents, $Ph_3P=CRR'$, which are used for the conversion of ketones into alkenes:

$$Ph_3P \xrightarrow{CH_3Br} Ph_3PCH_3^+Br^- \xrightarrow[-R''H.LiBr]{+LiR''} Ph_3P=CH_2 \xrightarrow{RR'CO} RR'C=CH_2 + Ph_3PO \qquad 31$$

The basic properties of organolithium compounds can also be exploited to generate **carbenes**, highly reactive very short-lived species of general formula CXX' (X and X' are monovalent atoms or groups), which formally contain divalent carbon; further reaction with an alkene, for example, is a means of generating a cyclopropane ring:

$$RCHCl_2 + LiBu \longrightarrow LiCl + BuH + \underset{\text{carbene intermediate}}{R\ddot{C}Cl} \xrightarrow{R'CH=CH_2} \underset{R'HC-CH_2}{\overset{RCCl}{\diagup\ \diagdown}} \qquad 32$$

Though organolithium and Grignard reagents are extremely useful synthetic reagents, alkyl zinc, ZnRR', alkyl mercury, RR'Hg, RHgI, or lithium dialkyl copper reagents, Li[CuRR'], are used when milder or non-basic conditions are required.

5.3 Organoboron and organoaluminium compounds

The organometallic chemistry of boron and aluminium is closely related, and is of great industrial and academic interest. Organoboron compounds undergo only slow hydrolysis because of the low polarity of the B—C bonds. The trialkylboron compounds, BR_3, are monomeric, and are therefore structurally different from the boron hydrides such as B_2H_6, B_4H_{10} and SeB_9H_{13}, which can be related to fragments of polyhedra with bridging hydrogens and direct boron–boron bonds.

SLC 4

At one time, boranes and carboranes (compounds containing carbon and boron) were of considerable interest as rocket propellants because their enthalpies of combustion exceed that of hydrocarbons. Organoaluminium compounds are used in a number of technical processes as sources of *carbanions*. Together with organoboron compounds they form a host of synthetic reagents for performing specific transformations in organic chemistry.

Organoboranes can be prepared in a number of ways (Figure 11).

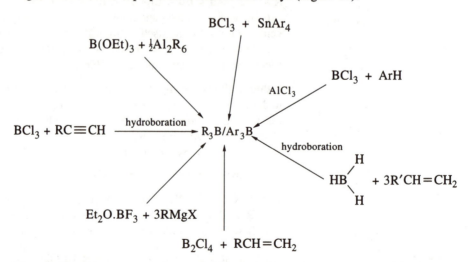

Figure 11 Preparative routes to trialkyl and triaryl boron compounds.

SLC 5

The most important preparative method is **hydroboration** (the insertion of alkenes or alkynes into a B—H or B—Cl bond), by which a number of different organoboron compounds can be synthesised. The rate of the hydroboration reaction falls as the boron atom becomes progressively more substituted, thereby allowing alkylboron hydrides to be isolated, as in reactions 33 and 34:

$$BH_3 \xrightarrow[thf]{C_2H_4} BEtH_2 \xrightarrow[thf]{C_2H_4} BEt_2H \xrightarrow[thf]{C_2H_4} BEt_3 \qquad 33$$

$$BH_3 \xrightarrow[thf]{Me_2C=CHMe} BH_2(MeCHCHMe_2) \xrightarrow[thf]{Me_2C=CHMe} BH(MeCHCHMe_2)_2 \qquad 34$$

The ethyl group (reaction 33) is small and so multiple substitution to form BEt$_3$ is not significantly sterically impeded. As a consequence, this means that BEtH$_2$ and BEt$_2$H are difficult to isolate. In reaction 34, however, both intermediate hydrides are bulky, so further reaction with Me$_2$C=CHMe is slow and the intermediates are isolable.

The American chemist H. C. Brown found that the hydroboration reaction enabled a range of organic products to be synthesised in a stereochemically controlled manner. The addition of HX to unsaturated hydrocarbons is a common reaction. The usual addition, in which the H of HX is added to the less-substituted carbon atom of a C=C bond, is called *Markownikoff addition*. Hydroboration takes place in the opposite sense, with the H of the B—H bond ending up on the *more*-substituted carbon atom of the C=C bond. This process is called **anti-Markownikoff addition**.

SLC 5

□   Suggest a reason based on bond polarities why the hydroboration reaction allows anti-Markownikoff addition?

■   The reason for this mode of addition is that the B—H bond is polarised B$^{\delta+}$—H$^{\delta-}$ because hydrogen is more electronegative than boron. (The orientation of attack is such that the most stable carbocation is formed; that is, the C$^{\delta+}$ is the more-substituted carbon). The hydroboration reaction involves a *cis* four-centre anti-Markownikoff addition of B—H to a C=C bond:

$$R-CH=CH_2 \xrightarrow{H-B} RC^{\delta+}\cdots CH_2^{\delta-} \longrightarrow RCH_2-CH_2-B \quad \text{35}$$

R = alkyl group, more bulky than an H atom

anti-Markownikoff addition

$$R-CH=CH_2 \xrightarrow{H^{\delta+}-Br^{\delta-}} RC^{\delta+}\cdots CH_2^{\delta-} \longrightarrow RHC(Br)-CH_3 \quad \text{36}$$

Markownikoff addition

Organoboranes can be used as intermediates for the formation of new carbon–carbon bonds by the migration of alkyl groups from boron to the carbon atoms of carbon monoxide, nitriles, alkynides, ketones and esters.

5.4 Organo derivatives of silicon and the heavier Group IV elements

The organo derivatives of the Group IV elements Si, Ge, Sn and Pb differ markedly from those of earlier Groups in that their metal–carbon bonds are less polar. Moreover, the tetra-alkyl and tetra-aryl molecules, MR$_4$, do not have available, vacant low-energy molecular orbitals, and so do not function as acceptors; nor do they associate through electron-deficient alkyl or aryl bridges. They are therefore less reactive to nucleophiles: most, for example, are unaffected at room temperature by air and water, though reactivity and thermal instability increase from silicon to lead, as their metal–carbon bonds become progressively longer, more polar and weaker. Their physical properties are those commonly associated with organic compounds: they are volatile liquids or low-melting solids, which dissolve in non-polar solvents.

Certain derivatives of silicon, tin and lead are of great industrial importance. Organosilicon compounds are produced on a large scale (>60 000 tonnes per annum) in the manufacture of silicone polymers, where their high thermal stability and low reactivity are extremely useful features. On the other hand, at the bottom of the Group, lead alkyls (PbMe$_4$ and especially PbEt$_4$) are currently produced in even greater tonnages for use as anti-knock additives for petrol, a role they can play because of their relatively low thermal stability and susceptibility to oxidation (in the engine cylinders, they form tiny lead oxide particles, which prevent pre-ignition). However, the increased use of lead-free petrol may soon eliminate this major use of organo-lead compounds. Between these extremes, organotin compounds find small-scale use as catalysts in polyurethane formation, and as polymer stabilisers (PVC, rubbers), since they are reactive enough to scavenge the substances that promote polymer decomposition, such as the initial products of ultraviolet irradiation or oxidation (radicals, HCl, unsaturated compounds). Alkyltin carboxylates also find widespread application as biocides, for example as agricultural fungicides, in wood and paint preservatives, for slime prevention in paper manufacture, and in veterinary medicines. Their use in marine anti-fouling paints has been subjected to careful monitoring in response to concern over their effect on higher marine life.

☐ Suggest a method for making tetraalkyl derivatives of Group IV elements.

■ In the laboratory, the most convenient methods of attaching organo groups to silicon, germanium, tin and lead use a Group IV halide and a Grignard, organolithium or organoaluminium reagent in a metathetical reaction, for example:

$$GeCl_4 + (LiMe)_4 \longrightarrow GeMe_4 + 4LiCl \qquad 37$$

When aluminium alkyls are used, the addition of a base is recommended to complex the $AlCl_3$ that is formed; otherwise, as in the case of tin, the $AlCl_3$ tends to stop the reaction at the stage of partial alkylation (formation of R_2MCl_2) by coordinating to its chlorine atoms and protecting the Group IV metal from further attack; for example

$$SnCl_4 + AlR_3 \longrightarrow R_2SnCl_2.AlCl_3 + R_3SnCl.AlCl_3, \text{ etc.} \qquad 38$$

Industrially, reactions between haloalkanes and the Group IV metals provide the preferred methods of synthesis. For example, the methylchlorosilanes, Me_nSiCl_{4-n}, used in the manufacture of silicones (see below), are prepared by passing chloromethane over silicon with a copper catalyst.

The so-called **double bond rule,** which states that elements of the third and higher rows in the Periodic Table do not form compounds containing $p\pi-p\pi$ multiple bonds, has challenged chemists to produce molecules containing $\diagdown\!\!\!\diagup\!\!\text{Si}\!=\!\text{Si}\diagdown\!\!\!\diagup$ and $\diagdown\!\!\!\diagup\text{Si}\!=\!CR_2$ components.

☐ Why should elements of the third and later rows avoid forming compounds containing multiple bonds, E=E or E=C?

■ The element–element $p\pi-p\pi$ multiple bond of third-row elements would be weak because of the poor orbital overlap of the 3p orbitals. This is caused by the long E—E σ bond for third- and higher-row elements, which have larger orbitals than second-row elements.

Silenes, containing Si=Si ($p\pi-p\pi$) bonds, have been synthesised at low temperature using photolysis techniques (Figure 12). However, steric protection of the vulnerable Si=Si multiple bond, for example by the bulky 2,4,6-trimethylbenzene groups shown in equation 39, is required for stability.

Compounds containing silicon–carbon double bonds have now been made and their X-ray crystal structures have been determined. Apart from steric protection, reactive multiple bonds can be stabilised by coordination to a transition-metal fragment.

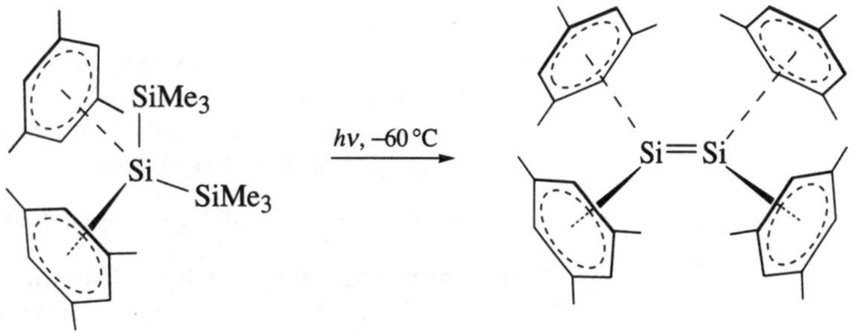

Figure 12 Synthesis of silenes by the low-temperature photolysis of $Si(C_6H_2Me_3)_2(SiMe_3)_2$.

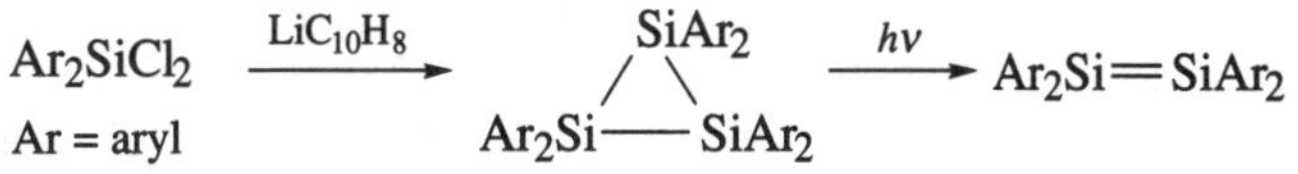

$$Ar_2SiCl_2 \xrightarrow{LiC_{10}H_8} \underset{Ar_2Si\text{———}SiAr_2}{\overset{SiAr_2}{\triangle}} \xrightarrow{h\nu} Ar_2Si=SiAr_2 \quad\quad 40$$
$Ar = aryl$

The much greater stability of Si—O bonds accounts for the remarkable industrial importance of **silicones**. Silicone polymers have skeletons of alternating silicon and oxygen atoms, the spare valences of the silicon atoms generally being occupied by terminally attached alkyl or aryl groups. The simplest examples have the formula $R_3Si(OSiR_2)_nOSiR_3$, in which the dialkylsilicon or, less commonly, diarylsilicon divalent chain units —SiR_2— are linked together by bridging oxygen atoms, the chains ending with monovalent units —SiR_3, as shown in structure **1**.

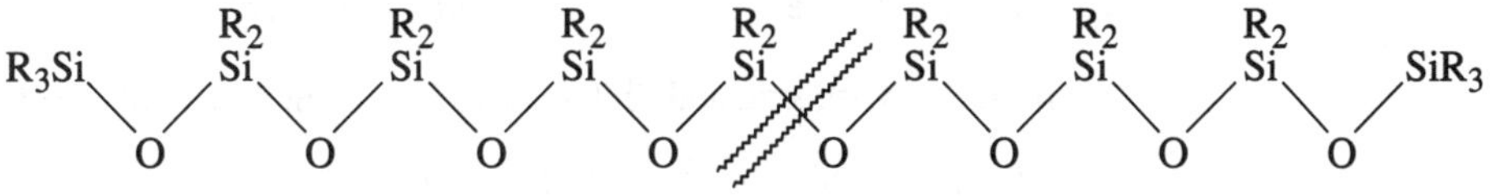

Cross-linking between such chains is achieved by incorporating trivalent units $\equiv SiR$ at intervals (structure **2**).

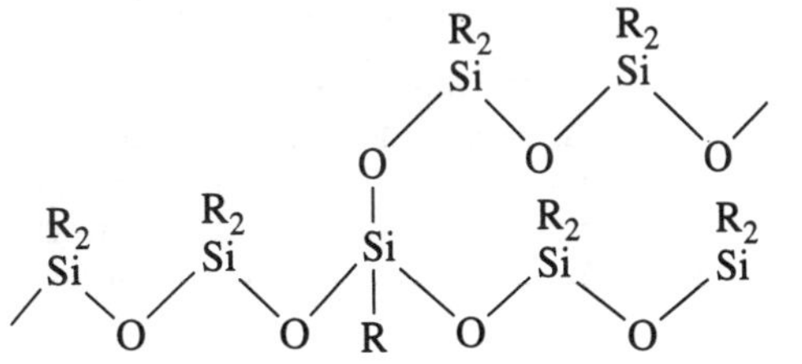

The number of such cross-links influences the bulk properties of the product. Linear polymers, $R_3Si(OSiR_2)_nOSiR_3$, are oils. Polymers with occasional cross-links have a rubbery texture, and polymers with many cross-links are relatively inelastic resins.

The particular respects in which silicone polymers are superior to polymers with organic backbones are: their high thermal stability and resistance to chemical attack; their electrical insulating properties, which persist at high temperatures (contrast to organic polymers, which tend to decompose to form sooty, electrically conducting materials); their low temperature coefficients of viscosity; and their water-repellent, non-stick surface properties. These have led to a wide variety of applications, including the use of silicone oils as lubricants that remain effective over wide temperature ranges, or as thinners for waxes (for cars, furniture), which can then be applied as aqueous suspensions. Resins are used for making non-stick surfaces, for example on tyre moulds and backing paper for self-adhesive materials. Silicones are also used for waterproofing fabrics, for stabilising polyurethane foams during polymerisation, and for surgical implants.

They are normally prepared from the chloro-organosilicon derivatives, R_nSiCl_{4-n} (particularly Me_3SiCl, Me_2SiCl_2, and $MeSiCl_3$). These compounds react violently and exothermically with water by the cleavage of the silicon–chlorine bonds to give $\equiv Si-OH$ and HCl, the silicon–carbon bonds remaining unaffected.

Hydrolysis of Me_3SiCl produces the disiloxane $(Me_3Si)_2O$; Me_2SiCl_2 gives cyclic siloxanes, $(Me_2SiO)_n$, and acyclic species, $HO(Me_2SiO)_nH$ (n = 3–9). In silicone oil manufacture these crude hydrolysis products are polymerised further using acid or alkali as a catalyst, the chain length being controlled by the proportion of the chain-ending species, $(Me_3Si)_2O$, added:

$$(Me_3Si)_2O + x(Me_2SiO)_n \longrightarrow Me_3Si(OSiMe_2)_{xn}OSiMe_3 \qquad 41$$

Resins are produced by the hydrolysis of Me_2SiCl_2 mixed with suitable proportions of $MeSiCl_3$. Their properties may be improved by incorporating other organic substituents, for example by including Ph_2SiCl_2 and $PhSiCl_3$ in the mixture of hydrolysed chlorosilanes.

We recommend that you now view Programme 2 of Videocassette 2, referring to Section 8 of the S343 Audiovision Booklet for guidance.

5.5 Organo derivatives of Group V and Group VI elements

The organometallic chemistry of Group V and Group VI elements has been the focus of much recent research. To date (1994) few applications from this field have been applied to commercial processes, especially when compared to the utility of organosilicon and organoboron compounds. However, a range of fundamentally new chemistry has been developed, some of which can be described as 'molecular architecture'. The development of this field has enabled a number of new compounds with unusual linkages to be formed; C—E=E—C, \>C=E and —C≡E (E = P, As, Sb, Bi, Se, Te), *in defiance of the double bond rule.*

Group V organometallic compounds exhibit a variety of structural types. This is feasible because two major oxidation states E^{III} and E^V exist. Variation in coordination number is therefore possible, leading to polymers, oligomers and Lewis base complexes. Although the C—E bond polarity decreases down the Group, the behaviour of organometallic compounds of P, As, Sb and Bi is broadly similar.

Numerous examples of complexes with the formulation R_5E are known. Although these are usually thermally and air stable, they are prone to hydrolysis reactions and the formation of 'ate' complexes, in which the Group V element bears a formal negative charge, and 'onium' ions, in which the Group V element bears a formal positive charge; for example

$$Me_5As + H_2O \longrightarrow [Me_4As]^+ \, OH^- + MeH \qquad 42$$

$$Ph_5As + BPh_3 \longrightarrow [Ph_4As]^+ \, [BPh_4]^- \qquad 43$$

$$Ph_5Sb + LiPh \longrightarrow Li^+ \, [SbPh_6]^- \qquad 44$$

Equations 42 and 43 show the formation of an onium ion, in tetramethylarsonium hydroxide and tetraphenylarsonium tetraphenyl borate, respectively. Reaction 44 shows the formation of the 'ate' complex lithium hexaphenylstibate.

ER_3 groups have been used extensively as ligands in transition-metal complexes. The E^{III} atom has a free lone pair, which acts as a Lewis base and also has π acceptor character (free d orbitals on the Group V element available for back donation, as shown in Figure 13).

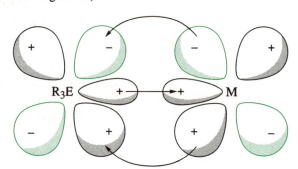

Figure 13 Schematic representation of metal–E bonding: σ donation of lone pair from E to metal, together with π back donation of electron density from filled metal d orbital to vacant d orbital on the E atom.

In general, ER_3 compounds are synthesised from reaction of the corresponding EX_3 (X = halogen) with a Grignard reagent, or by reacting the element with a haloalkane over a suitable catalyst.

Interestingly, Group V organometallic compounds possess the power of catenation, or self-linkage, through the Group V element.

The E—E bonds are weak (E_m(As—As) = 146 kJ mol^{-1}, E_m(Sb—Sb) = 120 kJ mol^{-1}), and become even weaker down a Group. Hence it wasn't until 1983 that the first well-characterised dibismuthane, Ph$_2$Bi—BiPh$_2$, was synthesised by the action of sodium in liquid ammonia on Ph$_2$BiCl. As indicated in the structure in reaction 45, this compound has bond angles that indicate almost pure p orbital interaction on the bismuth, which means that the lone pair is almost exclusively 6s in character:

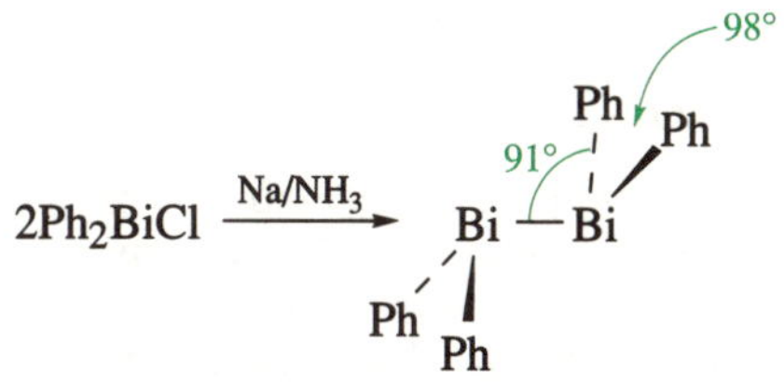

$$2Ph_2BiCl \xrightarrow{Na/NH_3} \underset{\substack{Ph \\ Ph}}{Bi}\!-\!\underset{\substack{Ph \\ Ph}}{Bi} \quad (91°, 98°) \qquad 45$$

Reactions of As^{III}R$_3$ and As^{V}R$_5$ units provide the building blocks to synthesise rings that contain just arsenic atoms (Figure 14.). Rings that contain only one type of atom in such compounds are called **homoleptic rings**.

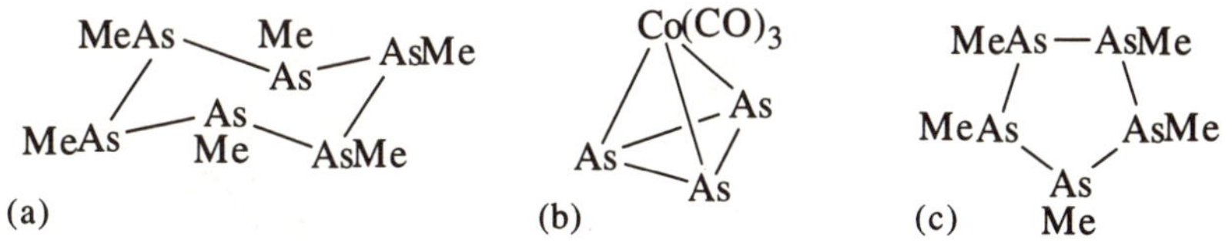

Figure 14 Arsenic organometallic rings: (a) (AsMe)$_6$ is analogous to cyclohexane; (b) [(Co(CO)$_3$As$_3$] has a tetrahedral shape, with the arsenic atoms in a three-membered ring; (c) (AsMe)$_5$ contains a five-membered ring of arsenic atoms.

It has also been possible to build up chains of arsenic atoms (Figure 15) by reduction of MeAsI$_2$ with Bu$_3$Sb:

$$n\text{MeAsI}_2 + n\text{Bu}_3\text{Sb} = \tfrac{1}{2}(\text{As}_2\text{Me}_2)_n + n\text{Bu}_3\text{SbI}_2 \qquad 46$$

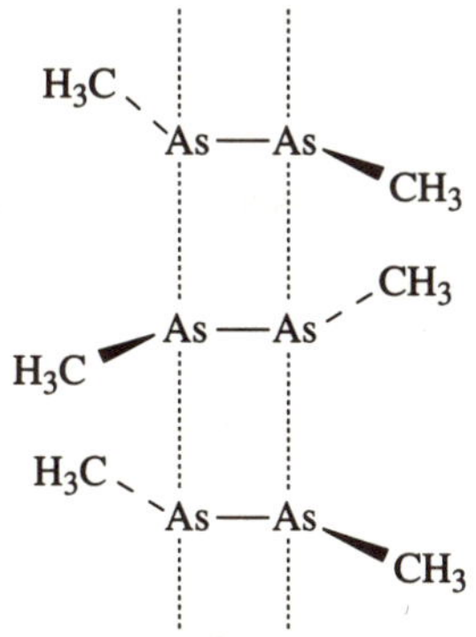

Figure 15 Chains of arsenic atoms linked together in sheets in (As$_2$Me$_2$)$_n$. Note that the arsenic atoms are stacked on top of one another, held by weak As—As bonding.

It should also be noted that heterocyclic organometallic complexes such as C$_5$H$_5$E and C$_4$EH$_3$RR' (Figure 16) (E = P, As, Sb) have been made. The six-membered rings are analogues of benzene, such that the bismuth analogue is a bismabenzene. The five-membered ring arsenic compounds rejoice in the name arsoles!

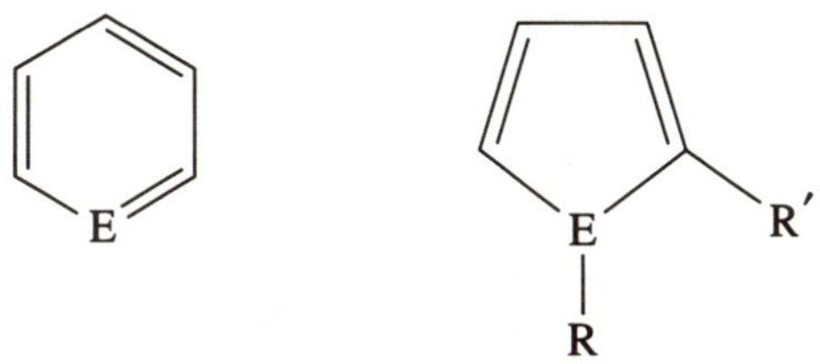

Figure 16 Organometallic arene compounds (E = P, As, Sb).

Recently a number of multiply bonded E=C, E=E and E≡C compounds have been synthesised. These compounds contain quite weak pπ–pπ bonds. Their stability is dependent on one of two factors — coordination to a transition metal or steric protection. Otherwise they are susceptible to attack by nucleophiles/electrophiles. An example of stabilisation of an As=As group by steric protection is shown in reaction 47:

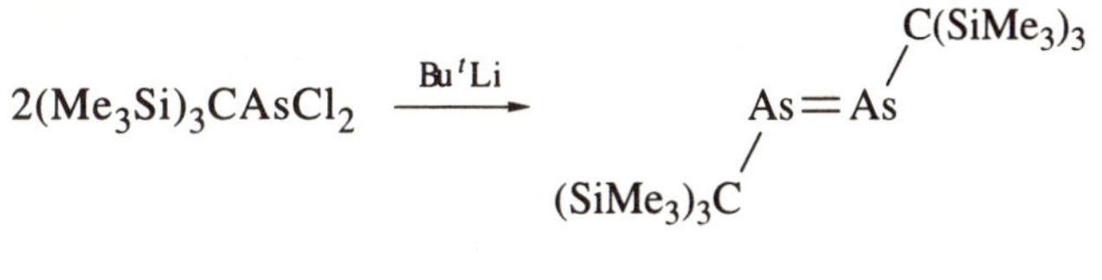

$$2(Me_3Si)_3CAsCl_2 \xrightarrow{Bu^tLi} (Me_3Si)_3C\text{–As}=\text{As–}C(SiMe_3)_3 \qquad 47$$

The first example of a molecule containing a $-C\equiv As$ fragment was reported in 1986 with the synthesis of $Bu^t-C\equiv As$; in this case steric protection was afforded by the large tertiary butyl group.

By contrast, *in situ* generation of a $-C\equiv As$ fragment has been achieved by coordination to a transition-metal centre:

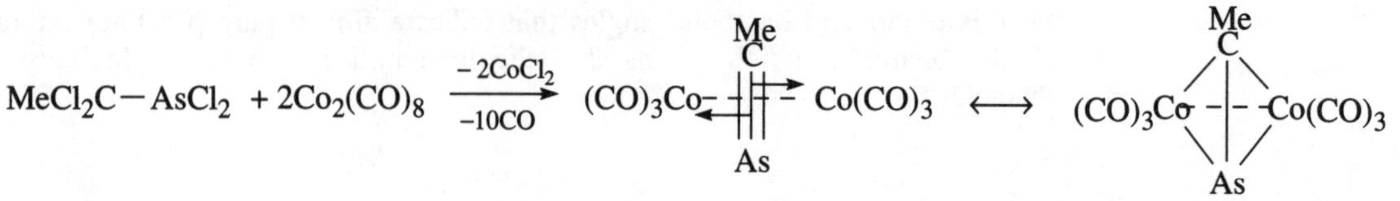

$$MeCl_2C-AsCl_2 + 2Co_2(CO)_8 \xrightarrow[-10CO]{-2CoCl_2} (CO)_3Co\underset{As}{\overset{Me\ C}{\rightleftarrows}}Co(CO)_3 \longleftrightarrow (CO)_3Co\underset{As}{\overset{Me\ C}{-}}Co(CO)_3 \quad \mathbf{48}$$

The phosphorus analogue, $Bu^t-C\equiv P$, has also been synthesised.

A number of compounds containing E=E multiple bonds have been made using the prerequisites of either steric protection or transition-metal coordination stabilisation. When E = P, the product is a diphosphine. A particularly stable *bis*(aryl) diphosphine is one in which stabilisation is achieved by attachment of the bulky supermesityl group, 2,4,6-*tris*(*t*-butyl)benzene, to the phosphorus atoms; see Figure 17.

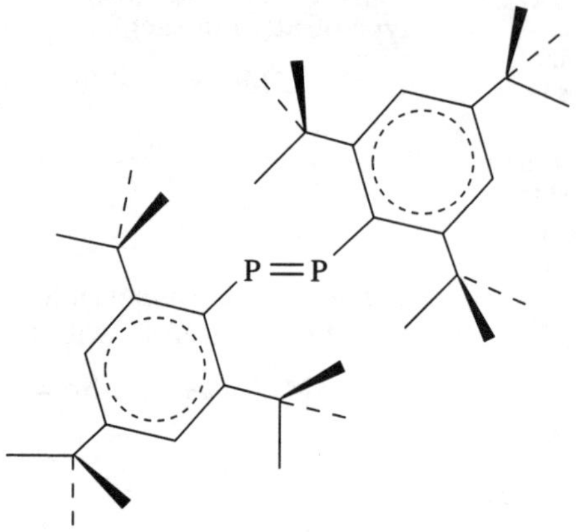

Figure 17 A *bis*(aryl)diphosphine containing the supermesityl group. The P—P distance of 203 pm is shorter than the P—P distance in the phosphorus tetramer, P_4 (224 pm), which is indicative of a P—P multiple bond in the *bis*(aryl)diphosphine.

Stabilisation of the $-As=As-$ and $-Sb=Sb-$ double bonds has also been achieved; the antimony analogue requires both steric protection and coordination to a transition metal:

$$(Me_3Si)_2CHSbCl_2 \xrightarrow{Na_2[Fe(CO)_4]} (Me_3Si)_2CH-\underset{\underset{Fe(CO)_4}{|}}{Sb}=Sb-CH(SiMe_3)_2 \quad \mathbf{49}$$

The multiple bonds in E=C, E=E and E≡C complexes can be rationalised as being of the pπ–pπ type (Figure 18).

In comparison to the organometallic chemistry of Group V, the organometallic chemistry of Group VI has less variety in structural types, although a number of interesting molecules has been synthesised. The organometallic chemistry of Group VI is restricted to the elements selenium and tellurium because oxygen and sulphur are non-metals and polonium is highly radioactive.

A major current emphasis in organometallic chemistry in which Group VI elements have been used is the use of volatile organometallic reagents in the preparation of thin films of semiconducting materials by **molecular organic chemical vapour deposition, MOCVD**. This procedure involves volatilising the organometallic compound into the gas phase under vacuum, and then passing the heated gas through a hot zone (typically 300–500 °C) in which the molecules are thermally decomposed. The decomposition causes deposition of the metal-containing fragments on to the reactor walls. This process can be exemplified by the formation of CdTe, a material with considerable promise as a semiconductor. Combination of $CdMe_2$ and $Te(C_2H_5)_2$ in the gas phase of an MOCVD reactor lays thin films of CdTe on the reactor walls.

Some metal alkyls and aryls can be made to form metal carbides by this technique; for example

$$TiR_4 \xrightarrow[MOCVD]{heat} TiC \quad \mathbf{50}$$

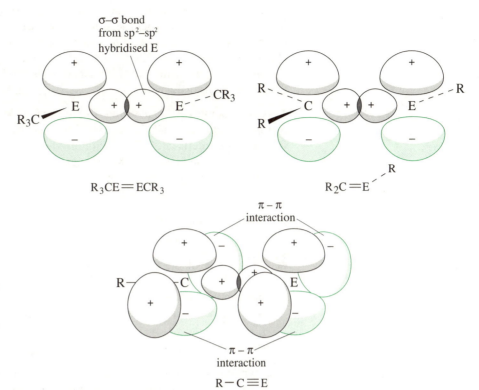

Figure 18 Bonding in E=E, C=E and C≡E complexes. For $R_3C-E=E-CR_3$ the E atoms can be considered to be sp^2 hybridised with a π bond formed by two p orbitals, one from each E atom. $R-C≡E$ has two mutually perpendicular π bonds formed from two of the p orbitals on E and two 2p orbitals from carbon. $R_2C=E-R$ has a π bond formed from two p orbitals, one based on carbon and the other based on the E atom.

The titanium carbide formed in this way can be used in drill bits for the oil and tool-cutting industries.

Organoselenium compounds are critical to the functioning of the human body. Ironically, a daily intake of organoselenium in food exceeding 1 μg is toxic, whereas less than 0.2 μg per day causes liver malfunction and visual impairment. Organoselenium is found in the body as a component of glutathione peroxidase and in selenomethionine.

5.6 *Monohapto* derivatives of the transition metals

Most transition metals have been shown to be able to coordinate *monohapto* hydrocarbon ligands, though complexes containing such ligands are less common than complexes containing unsaturated *dihapto* or *polyhapto* ligands. Relatively few simple alkyls or aryls, MR_n, for example, are known, and those that have been isolated have lower thermal stability and higher reactivity than analogous derivatives of main-Group metals. One reason for this is that transition metals, having incomplete d shells of electrons, have more orbitals available for bonding, but they cannot make effective use of all of them if the ligands themselves furnish only one orbital and one electron each (as simple alkyl or aryl groups do) for metal–carbon bonding. For example, pairing all six electrons of a Group VIA metal atom (Cr, Mo or W) with an electron each from six methyl groups, as in $[MMe_6]$, would leave three orbitals on the metal unused, and these would be available for reaction with nucleophiles or for rearrangement reactions.

Various categories of organic group have been shown to coordinate *monohapto* to transition metals. Those that form metal–carbon single (σ) bonds include simple alkyl and aryl groups, and their fluorocarbon analogues, also acyl, alkenyl and alkynyl groups. Examples of compounds containing such ligands include the manganese acyl carbonyl complex $[Mn(COCH_3)(CO)_5]$ and the iron alkenyl carbonyl complex $[Fe(CH=CH_2)(\eta^5-C_5H_5)(CO)_2]$. In addition to these examples, there are other ligands that form multiple bonds to metals, as in carbene, M=CRR′, and carbyne, M≡CR, complexes.

A typical decomposition reaction of alkyl derivatives with β-attached hydrogen atoms involves the elimination of an alkene:

$$RCH_2CH_2ML_n \longrightarrow RCH=CH_2 + MHL_n$$

In metal aryls, such a decomposition process does not normally occur because β-hydrogen extraction would lead to formation of a strained triple bond, which is a very high-energy process. For metals later in the transition series, transfer of electronic charge from filled metal orbitals into ligand π^* orbitals is a further possibility. As a result, transition-metal aryls tend to be thermally more stable than their alkyl analogues.

Alkynyl complexes, containing the grouping M—C≡CR, may be yet more thermally stable than alkyls or aryls. The nickel complex $[Ni(C≡CPh)_2(PEt_3)_2]$, for example, melts at 150 °C without decomposition. Like its isoelectronic analogues CN^- and CO, the ethynide ligand $C≡CH^-$ may be expected to have both σ-donor and π-acceptor character.

☐ Why are the $v(C≡C)$ stretching frequencies of metal alkynides, R—C≡CML_n, some 150 cm^{-1} lower than those of related alkynes, RC≡CR'?

■ Weakening of the C≡C bond in the transition-metal complexes arises from back donation of electrons from metal d orbitals to the empty antibonding π^* orbitals of the ethyne ligand. In such a process the carbon–carbon bond order is decreased and the i.r. stretching frequency is lowered.

Metal–carbon bonds in these alkynides are clearly stronger than single σ bonds.

The methods of synthesis and types of reaction of transition-metal alkyls and aryls are like those already described for main-Group compounds; for example

$[PtCl_2(PEt_3)_2]$
- MeMgI → $[PtIMe(PEt_3)_2]$
- 2MeLi → $[PtMe_2(PEt_3)_2]$ **52**
- 2RC≡CLi → $[Pt(C≡CR)_2(PEt_3)_2]$

$CrCl_3(thf)_x + 3PhMgBr \longrightarrow CrPh_3(thf)_x + 3MgBrCl$ **53**

A particularly important property of transition-metal alkyls and aryls with regard to their catalytic role is their capacity to undergo **insertion reactions**:

$$R-ML_n + X \longrightarrow R-X-ML_n \quad\quad\quad \mathbf{54}$$

where X = CO, SO_2, $F_2C=CF_2$, etc. In the carbonylation reaction, the metal alkyl or aryl is converted into the corresponding acyl derivative, $[M(RCO)L_n]$. Studies of carbonylation reactions of metal alkyl–carbonyl complexes, $[MR(CO)_n]$, in the presence of isotopically labelled free carbon monoxide have shown that it is a coordinated CO group that inserts into the metal–carbon bond by an intramolecular shift, and not a CO molecule from the gas phase:

$$L_nM-R \atop \underset{O}{\underset{\|\|}{C}} \longrightarrow L_nM-O=C-R \xrightarrow{CO} L_nM-CO,\ O=C-R \quad\quad \mathbf{55}$$

The acyl complexes formed by such insertion reactions can themselves be converted into carbene complexes containing RC(OR´) ligands by alkylation of the acyl oxygen atom; for example

$$[Cr(CO)_6] + LiR \longrightarrow Li^+\left[(OC)_5CrC\underset{O}{\overset{R}{\diagup}}\right]^- \xrightarrow{R'_3O^+} \left[(OC)_5CrC\underset{OR'}{\overset{R}{\diagup}}\right] \quad \mathbf{56}$$

5.7 Stability of *monohapto* organometallic compounds

A common source of instability of *monohapto* derivatives arises from the process

$$L_nM-\underset{\underset{H}{|}}{\overset{\overset{H}{|}}{C_\alpha}}-\underset{\underset{H}{|}}{\overset{\overset{H}{|}}{C_\beta}}-R \longrightarrow L_nM-H + \overset{H}{\underset{H}{>}}C=C\overset{R}{\underset{H}{<}} \qquad 57$$

where L_n represents n ligands L attached to the metal.

☐ This type of fragmentation process is facilitated by three conditions. Write down what you think these are.

■ Fragmentation will occur when:
   (a) the metal is able to increase its coordination number by one;
   (b) a substituent attached to the β-carbon atom is readily transferred to the metal centre;
   (c) the β-carbon atom is able to form a multiple bond with the α-carbon atom.

These features may seem at first to restrict the range of transition-metal *monohapto* complexes, but there are various ways in which this fragmentation process may be prevented. It is useful to consider each feature in turn because the properties highlighted help us to appreciate the role of the metal in the various stages of metal-catalysed reactions, many of which have industrial importance. We shall consider this point in more detail in Section 8.7.

(a) A metal is said to be **coordinatively unsaturated** if it can increase its coordination number — that is, the number of atoms directly bonded to it. The presence of such metals facilitates the fragmentation process. In some complexes containing a coordinatively saturated metal atom, especially those with bulky ligands, there may be a spontaneous loss of ligands when the complex dissolves, leaving the metal in a coordinatively unsaturated state. This type of behaviour is particularly common for phosphine complexes; for example, triphenylphosphine dissociates from platinum *tetrakis*(triphenylphosphine):

$$[Pt(PPh_3)_4] \longrightarrow [Pt(PPh_3)_3] + PPh_3 \qquad 58$$

Metal alkyls that remain coordinatively saturated are expected to be more stable than alkyls that readily lose a ligand. Coordinative saturation may be maintained using less-bulky ligand groups.

(b) An alternative approach to the production of stable alkyls is to take account of the possible fragmentation of the alkyl group and to design the alkyl so that transfer of a fragment group to the metal is no longer feasible. Since the transfer of groups other than hydrogen from the β-carbon atom to the metal is not very likely, alkyl groups having no β-hydrogens should produce relatively stable derivatives.

☐ Suggest substituted alkyl groups that you consider would promote the formation of more-stable derivatives.

■ Groups with fluoro and methyl substituents on the β-carbon atom are suitable, for example $-CH_2CMe_3$ and $-CH_2CF_3$ groups. Alternatively, the β-carbon may be part of a phenyl group (for example $-CH_2C_6H_5$), and a β-carbon is avoided completely if the alkyl group is methyl.

(c) Multiple bonding between the α- and β-skeletal atoms may be prevented by the presence of a β-atom having little or no tendency to π-bond. For example, β-silicon atoms do not readily participate in multiple bonds (double bond rule), and when they have substituent methyl groups many transition-metal alkyls can be synthesised; for example

$$\text{CrCl}_3 + \text{LiCH}_2\text{SiMe}_3 \xrightarrow[\text{thf}]{-\text{LiCl}} \text{Cr(CH}_2\text{SiMe}_3)_3 \xrightleftharpoons[]{\text{thf}} \text{Cr(CH}_2\text{SiMe}_3)_4 + \text{other chromium-containing materials} \qquad 59$$

Here the chromium(III) and chromium(IV) compounds are in equilibrium with each other in tetrahydrofuran. The corresponding simple alkyl derivatives of chromium are very unstable by comparison, and, except for methyl derivatives, cannot be isolated. It would be misleading to infer that complexes of transition metals with simple alkyl ligands containing β-hydrogen atoms are therefore unknown or very rare. Indeed, a large number are known, but they all have at least one feature in common; they all contain strong-field ligands, or involve metals having certain electronic configurations (d^3 and low-spin d^6).

☐ From the following list of ligands, select those generally regarded to bond strongly to transition metals:

CN^-, F^-, NH_3, CO, PR_3, H_2O, $\text{H}_2\text{NCH}_2\text{COO}^-$, 2,2′-bipyridyl

What general features do these strong-field ligands have?

■ CN^-, CO, PR_3, 2,2′-bipyridyl

The strong-field ligands are those with empty orbitals of π symmetry (d atomic orbitals or π* molecular orbitals), which are able to increase the ligand-field splitting energy, Δ, by bonding with metal d-type orbitals of the same symmetry (for example, the t_{2g} orbitals for an octahedral complex). Ligands such as CO, CN^-, and 2,2′-bipyridyl have suitable π* orbitals available, whereas alkyl phosphines and arylphosphines (PR_3) may use d orbitals. Each of these ligands is able to 'back-bond' in a synergic manner with the metal and hence strengthen the metal–ligand bond.

Thus, the stability of the majority of transition-metal complexes containing *monohapto* ligands depends on the presence of strong-field ligands such as carbonyl, phosphine or RNC groups. For example, $[\text{Mn}(\text{C}_2\text{H}_5)(\text{CO})_5]$ is thermally stable at temperatures well above ambient, whereas complexes with weaker-field ligands in place of carbon monoxide are less stable. The presence of strong-field π-bonding ligands does not necessarily lead to strong metal–alkyl bonds, but the bond enthalpy of other groups to the metal is generally increased, ensuring that the metal is coordinatively saturated. There is therefore a decreased tendency for decomposition to a coordinatively unsaturated complex.

It is relevant here to mention the vitamin B_{12} coenzyme, which was the first naturally occurring transition-metal compound with a metal–carbon σ bond to be recognised. In this complex, cobalt(III) is strongly N-bonded to a porphyrin-type ring system and is also N-bonded to a 5,6-dimethylbenzimidazole derivative (L, Figure 19). The sixth coordination position is occupied by a deoxyadenosyl group (R) bonded through carbon. The function of this coenzyme, an anti-anaemia factor, is concerned with the transfer of methyl groups.

Figure 19 Features of the structure of the vitamin B_{12} coenzyme.

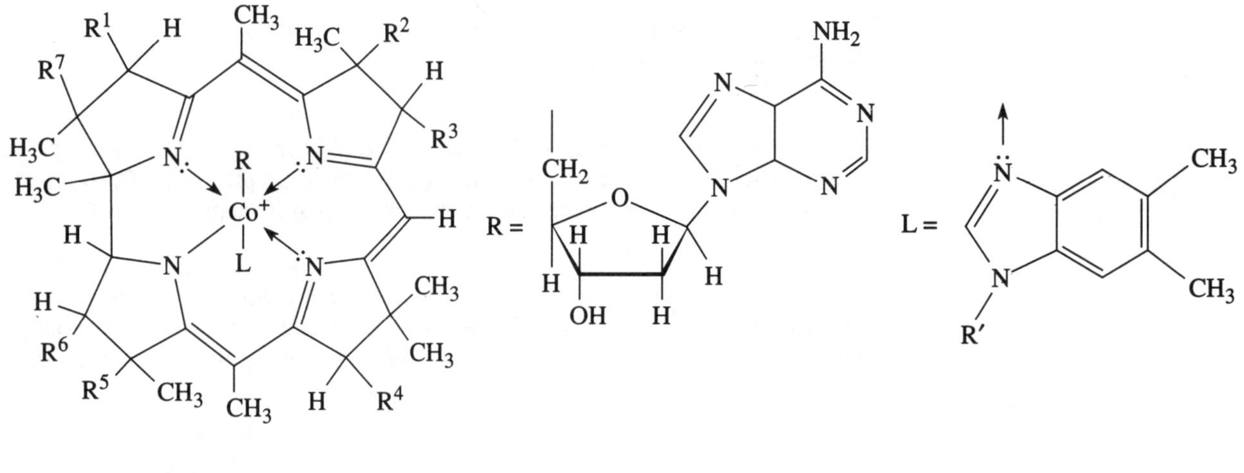

SAQ 4 Select from the following list of ligands the alkyl groups that are likely to form stable transition-metal complexes:

CH_3, $CH_2CH_2CH_3$, $CH_2CH_2SiMe_3$, CH_2SiHMe_2, $CH_2C_6H_{11}$, $CH_2C_6H_5$, CH_2CH_2SMe

SAQ 5 List the factors that are important in rationalising the stability of hexamethyltungsten(VI), $[W(CH_3)_6]$, towards fragmentation. Making reference to the chemistry of *monohapto* derivatives of the main-Group metals, suggest a synthetic route to $[W(CH_3)_6]$, using WCl_6 as the starting material.

5.8 Metal carbenes and carbynes

One facet of coordination to a transition metal is its ability to stabilise very reactive fragments. An example of this is the existence of carbene complexes. Free carbenes, $\cdot\ddot{C}R_2$, are electron deficient and have short lives in the free state; their high reactivity makes them useful in organic synthesis. Metal carbenes can be isolated, and can be classified into two types — Fischer carbenes (Figure 20a) and Schrock carbenes (Figure 20b). Although superficially very similar, their reactivity is quite different.

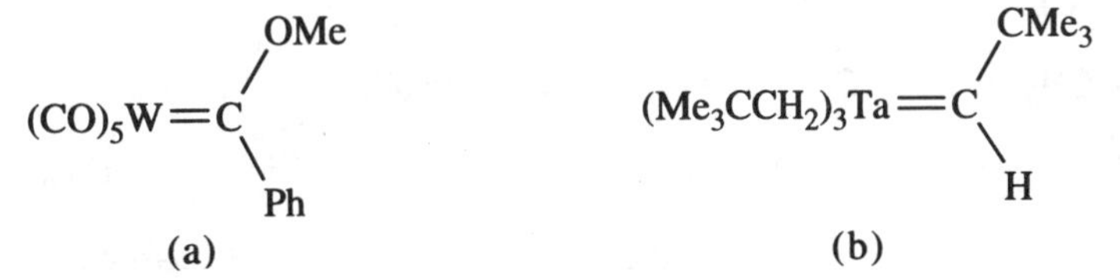

Figure 20 (a) An example of a Fischer-type carbene; (b) an example of a Schrock-type carbene.

Fischer-type carbenes contain metals in formally low oxidation states, and they can be used as a source of the neutral carbene. On heating, they give alkenes, via dimerisation of the carbene. As the electron density at the carbene carbon is low, the carbene carbon atom is electrophilic, making it liable to nucleophilic attack. One recent example of a Fischer-type carbene used as a reagent is in the synthesis of vitamin K (Figure 21).

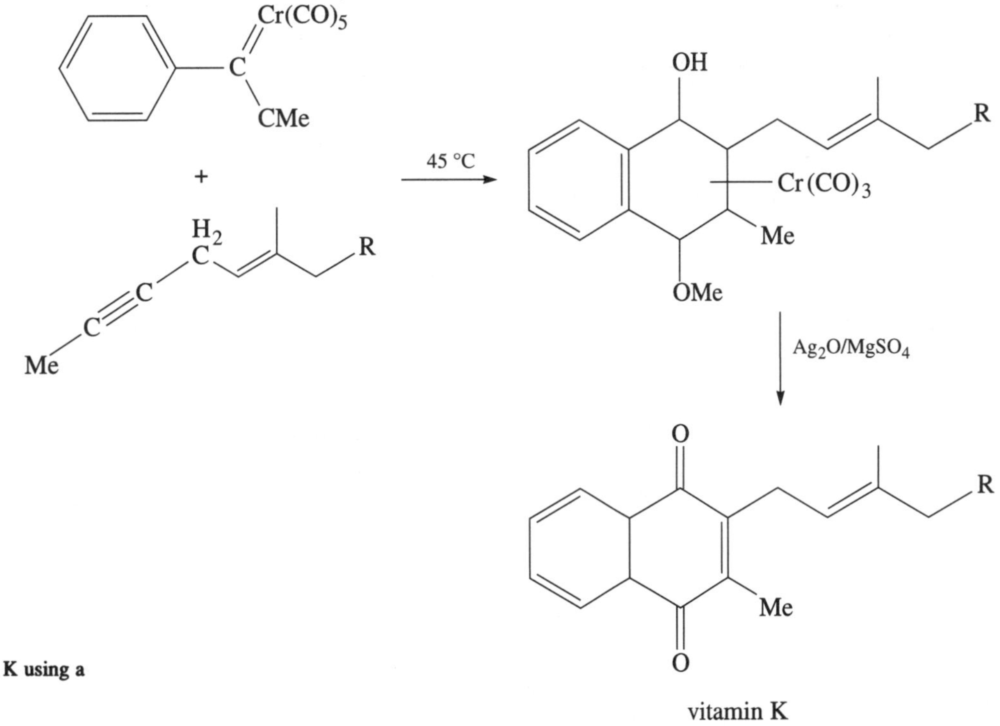

Figure 21 Synthesis of vitamin K using a Fischer-type carbene.

Schrock-type carbenes contain metals in formally high oxidation states. Indeed the compound can be considered to be quite polarised — that is, $R_3\overset{\delta+}{Ta}-\overset{\delta-}{C}R_2$. Therefore the carbon atom of the Schrock-type carbene has high electron density and is nucleophilic, which is the opposite situation to that found for Fischer carbenes. Carbenes of both types find application in organic synthesis and in catalyst-mediated reactions (as will be illustrated in Section 8).

Carbene complexes may be formed by α-hydrogen elimination. The tendency for metal alkyls with β-hydrogen atoms to form metal hydrides and eliminate alkenes has already been discussed. However, when the metal is in a high oxidation state, any attached alkyl groups will have relatively acidic hydrogen atoms at the α-position. As a consequence, deprotonation may occur on treatment with a strong base, via an intramolecular route. For example, the tantalum alkylidene complex shown as structure (c) in Figure 1 is formed as follows:

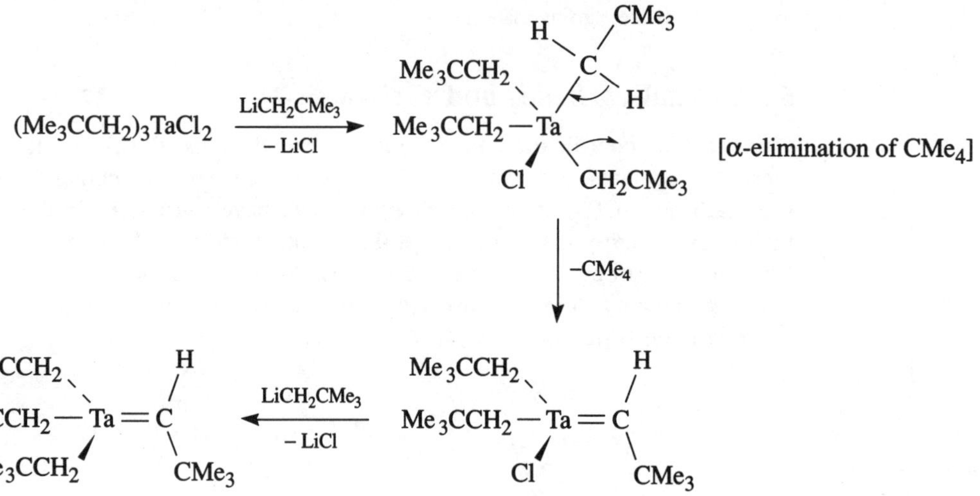

$$(Me_3CCH_2)_3TaCl_2 \xrightarrow[-LiCl]{LiCH_2CMe_3} \text{intermediate} \quad [\alpha\text{-elimination of } CMe_4]$$

60

$$\xrightarrow{-CMe_4} Me_3CCH_2-Ta=C(H)(CMe_3) \text{ with } Cl, CH_2CMe_3 \xrightarrow[-LiCl]{LiCH_2CMe_3} (Me_3CCH_2)_3Ta=C(H)(CMe_3)$$

Other routes to low oxidation-state metal carbenes involve attack on a metal-bound carbon monoxide ligand by an alkyllithium or amide:

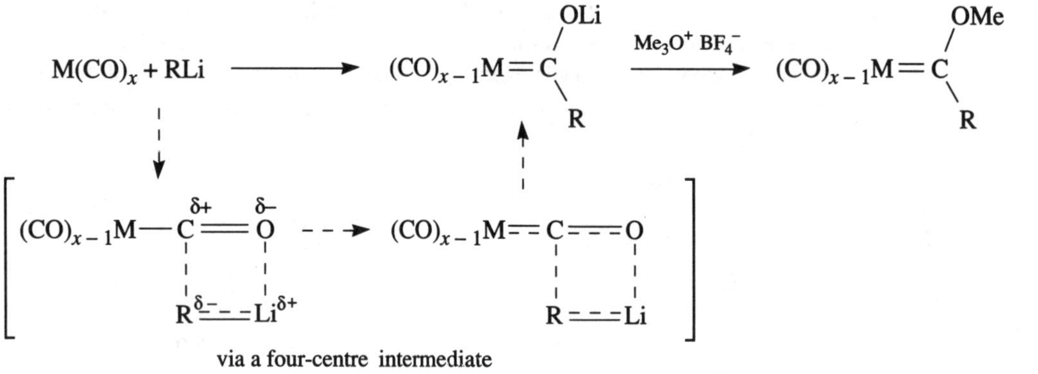

$$M(CO)_x + RLi \longrightarrow (CO)_{x-1}M=C(OLi)(R) \xrightarrow{Me_3O^+ BF_4^-} (CO)_{x-1}M=C(OMe)(R)$$

61

$$\left[(CO)_{x-1}M-\overset{\delta+}{C}=\overset{\delta-}{O} \text{ with } R^{\delta-}---Li^{\delta+} \quad \longrightarrow \quad (CO)_{x-1}M=C=O \text{ with } R---Li \right]$$

via a four-centre intermediate

$$M(CO)_x + LiNMe_2 \longrightarrow (CO)_{x-1}M=C(OLi)(NMe_2) \xrightarrow{Et_3O^+ BF_4^-} (CO)_{x-1}M=C(OEt)(NMe_2)$$

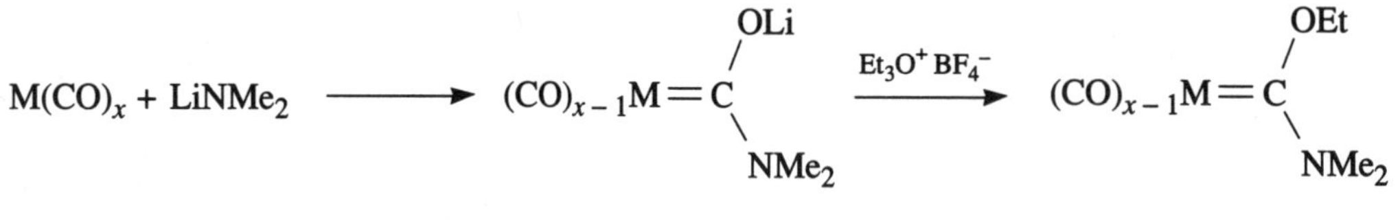

62

Transition metals can also bind to carbyne ligands $\cdot\overset{..}{C}-R$. The carbyne complex is the final product of a multistep reaction between a metal carbene and a boron halide (the mechanism is beyond the scope of this Course):

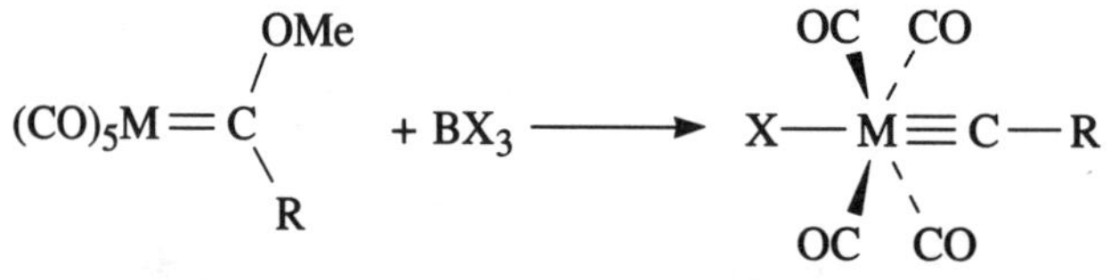

$$(CO)_5M=C(OMe)(R) + BX_3 \longrightarrow X-M\equiv C-R \text{ (with 4 CO cis)}$$

63

substitution is always *cis*

Another, and perhaps the mildest method of forming a metal carbyne, is to react a dimolybdenum hexa-alkoxide with an alkyne via a metathesis reaction:

$$(RO)_3Mo\equiv Mo(OR)_3 + R'-C\equiv C-R' \longrightarrow 2(RO)_3Mo\equiv C-R'$$

64

The M−C bond in the carbyne is formally a triple bond analogous to the one in ethyne with which you are more familiar. However, the metal normally uses d orbitals to form the 'π' component of the bond (see Section 6).

As would be expected, the metal–carbon distance varies with the bond order, just as for hydrocarbons; see, for example, Figure 22. Note, however, that variations at the metal cause the metal–carbon lengths to vary across the Periodic Table.

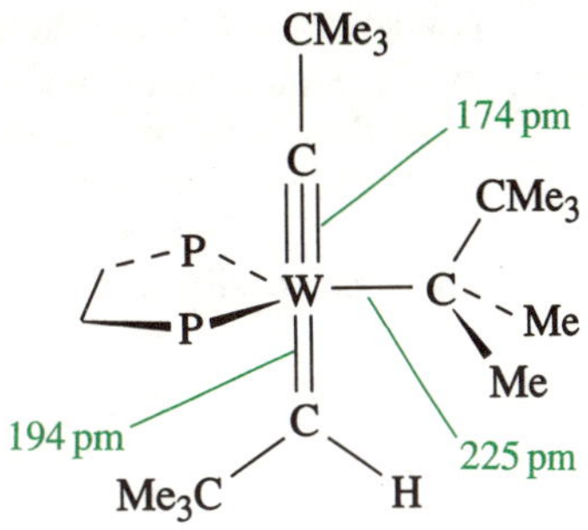

Figure 22 An example of an organometallic compound that contains M—C, M=C and M≡C bonds.

☐ Why is it that multiple bonds of the type E=E and E=C (E = As, Sb, Bi) are difficult to form?

■ The E=C bonds require either steric protection or transition-metal coordination stabilisation in order to be isolated.

SAQ 6 Determine the oxidation state of the complexes **a** to **h**. Use this information to help you classify the complexes as Fischer- or Schrock-type carbenes, and hence state which are susceptible to nucleophilic and which to electrophilic attack. (*Hint* Assume that the carbene ligand carries a double negative charge and that alkyl and ethoxide (OEt) carry a single negative charge when assessing the oxidation state of the metals.)

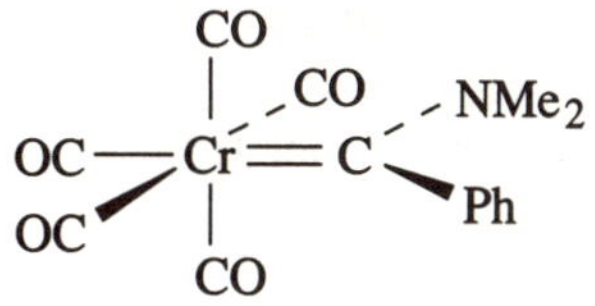

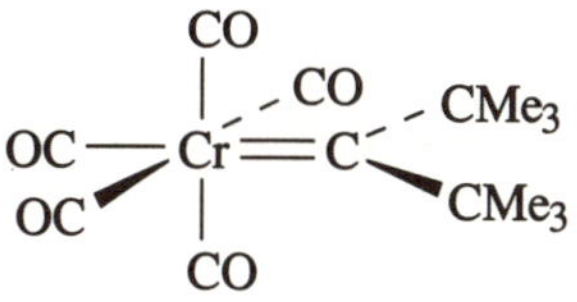

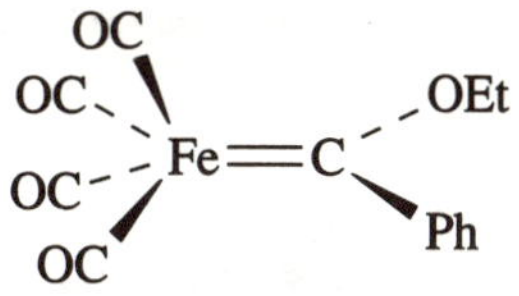

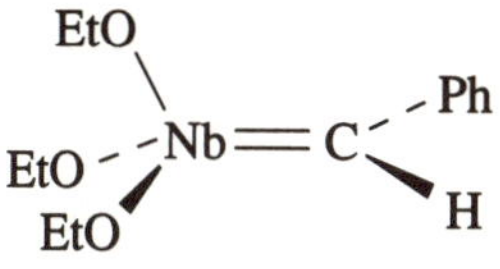

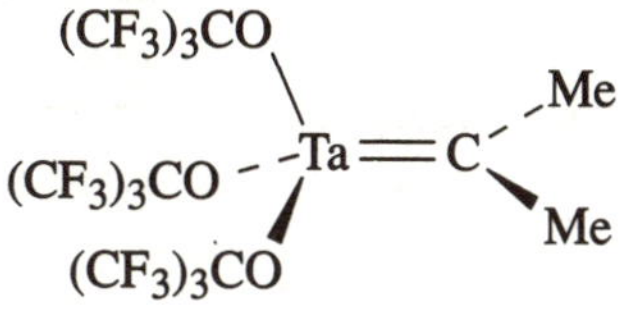

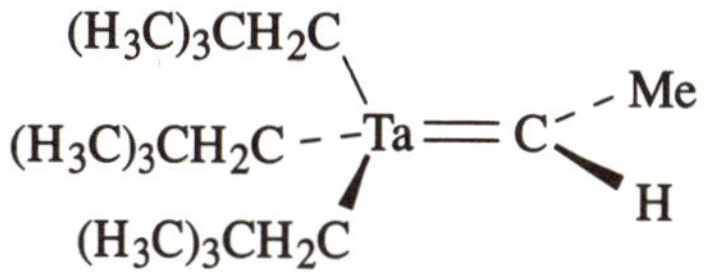

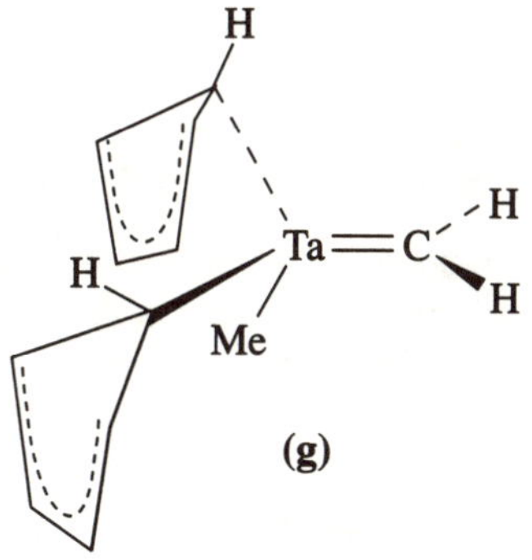

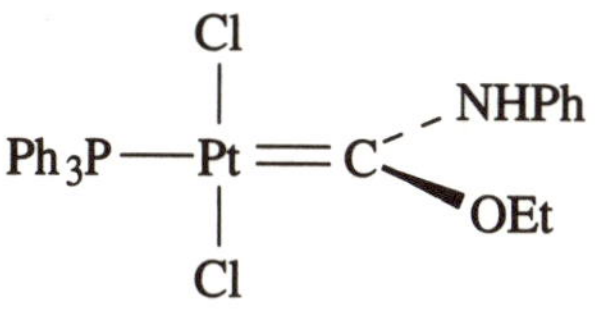

5.9 Agostic interactions

One fascinating insight into how particular organometallic transformations can occur was gained by the discovery that a C—H bond of a ligand may combine with a central metal atom to form a three-centre two-electron bond, M—H—C. Such bonds strongly resemble the *B—H—B bridges in boron chemistry* you encountered in a Second Level Course. M—H—C bonds, however, are somewhat less symmetrical; they have been called **agostic**† bonds (Figure 23). Agostic interactions may contribute to our understanding of C—H activation at transition-metal centres. Agostic C—H—M bridges have been detected mainly in complexes in which C—H bonds are positioned β to a coordinated alkyl or conjugated diene unit.

SLC 6

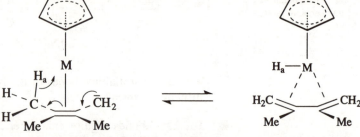

Figure 23 Agostic hydrogen (H_a). The bonding of the agostic hydrogen H_a can be represented by the representations shown, since its properties are intermediate between those of a hydrogen atom in a methyl group of a metal alkenyl (left) and a hydrogen atom coordinated to the metal in a metal hydride (right).

The evidence for agostic interactions has often been disputed because X-ray crystallography, which is used to determine atomic positions, is of limited use. This is because hydrogen atoms scatter X-rays so weakly that they cannot normally be detected when they are close to heavy-metal atoms. However, four main analytical methods can be employed to detect or infer the presence of agostic bonds:

(i) neutron diffraction (similar to X-ray diffraction but is able to locate hydrogen nuclei);

(ii) v(C—H) stretching vibration, which imay be in the range $2\,700-2\,300\,\text{cm}^{-1}$, since the C—H bond is weakened by the agostic interaction;

(iii) ^{1}H n.m.r. chemical shifts of agostic hydrogen, which are often less than 0 p.p.m. and are somewhat similar to those observed in metal hydrides (Block 5, Figure 12);

(iv) the $^1J(^1H, ^{13}C)$ n.m.r. coupling constants for agostic hydrogens are small (75–100 Hz).

The agostic interaction may be a prelude to complete C—H activation, in which the carbon–hydrogen bond is broken. In an agostic interaction the C—H bond has been 'activated' by the transition-metal complex and this may have important consequences in catalysis.

☐ Why do agostic hydrogen atoms generally have small $^1J(^1H, ^{13}C)$ coupling constants and chemical shifts less than 0 p.p.m.?

■ The agostic interaction lengthens and weakens the C—H bond. Coupling constants are transmitted via electrons in s orbitals; the poorer the overlap of carbon and hydrogen s orbitals (due to the formation of an agostic bond between the hydrogen and the metal), the smaller will be the $^1J(^1H, ^{13}C)$ coupling constant. The agostic hydrogen can be considered to be partially bound to the metal and hence is shielded from the external magnetic field by the metal electrons, especially those in d orbitals. Hence agostic hydrogens are similar to those in metal hydrides in that they normally have low-frequency chemical shifts.

5.10 Summary of Section 5

1 Grignard reagents, prepared by the addition of RX to magnesium in ethereal solvents, can be used to effect many organic syntheses.

2 Organolithium compounds can be synthesised from haloalkanes and lithium metal in hydrocarbon solvents. Ethers react with LiR compounds because they tend to be stronger bases than Grignard reagents.

† The name is derived from the awkward bent position of an arm in a Greek statue.

3 Organoboron compounds, unlike organolithium and organomagnesium compounds, tend to be hydrolytically stable. The hydroboration of alkenes in ethereal solvents is the first step in organic syntheses that use organoboranes as reactive intermediates.

4 Among organo derivatives of the main-Group metals, those of the Group IV elements most closely resemble typical organic compounds in reactivity. In compounds MR_nX_{4-n}, the centre of reactivity is the M—X bond rather than the M—C bond, which is relatively non-polar.

5 Although, in recent years, compounds with multiple bonds to silicon have been isolated, most organosilicon species are singly bonded.

6 Organometallic C=E and E=E bonds can be stabilised by steric protection using bulky substituents and by coordination to a transition metal.

7 Molecular organic chemical vapour deposition (MOCVD) using organometallic agents can be used to deposit thin films of new materials such as CdTe and TiC.

8 Among *monohapto* organo derivatives of transition metals, the least stable include those able to decompose by β-hydrogen elimination:

$$ML_nCH_2CH_2R \longrightarrow ML_nH + CH_2{=}CHR \qquad 65$$

9 This mode of decomposition is not possible when the metal is attached to groups such as CH_3, CF_3 or C≡CMe, which are β-hydrogen stabilised.

10 Apart from the decomposition route above, *monohapto* transition-metal derivatives can undergo insertion reactions.

11 Metal carbenes are of Schrock or Fischer types, and have different reactivity to electrophiles/nucleophiles. Fischer carbenes contain metals in a low oxidation state and Schrock carbenes contain metals in a high oxidation state. Fischer carbenes are susceptible to nucleophilic attack, whereas Schrock carbenes are susceptible to electrophilic attack.

12 Agostic interactions involve formation of a three-centre C—H---M bond. The C—H bond is lengthened in this interaction; its i.r. stretching frequency and $^1J(^1H, ^{13}C)$ coupling constant are less than for a normal C—H bond.

6 ACYCLIC *POLYHAPTO* SYSTEMS

6.1 Introduction

Attachment of an alkene to a transition metal has been known since the discovery of Zeise's salt in 1831 (Section 2.1). When ethene is shaken for several days with an aqueous hydrochloric acid solution of $K_2[PtCl_4]$, it yields Zeise's salt, potassium ethenetrichloroplatinate(II), $K[(PtCl_3(C_2H_4)].H_2O$, on crystallisation. The way in which the ethene molecule is bonded in the complex remained unclear for over a century, and it was only in 1954 that the crystal structure was determined by X-ray methods. It shows the usual square-planar geometry about platinum (Figure 24), with three of the four coordination sites occupied by chlorine atoms. The water molecule is not involved in direct attachment to the metal, but is accommodated in the lattice as a solvent molecule of crystallisation. The ethene occupies the fourth coordination position, the two carbon atoms being equidistant from the metal, and the carbon–carbon bond is at right-angles to the $PtCl_3$ plane.

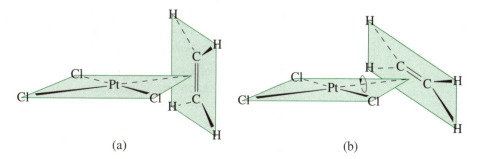

Figure 24 Zeise's salt, $K[PtCl_3(C_2H_4)].H_2O$: (a) structure of the anion; (b) an alternative conformation obtained by alkene rotation about the metal–alkene axis.

In the crystal, there is apparently a preference for the arrangement of atoms shown in Figure 24a over the alternatives obtained by rotation of the ethene about the platinum–ethene axis (one of which is represented in Figure 24b), because repulsions between ethene hydrogen atoms and adjacent chlorines are minimised in conformation (a). Consider now the position of the metal in relation to ethene. The platinum is located to the left of the plane of the alkene, symmetrically between the two carbon atoms.

☐ What is the *hapticity* of ethene in Zeise's salt?

■ Ethene is bonded to the platinum as a *dihapto* ligand, η^2-C_2H_4.

☐ Considering that platinum has vacant bonding orbitals, what feature of the electronic structure of ethene do you think would encourage the metal to take up this position?

■ Transition-metal cations are invariably Lewis acids, and are attracted towards molecules and ions having electron density available for bonding. Ethene has a π bond arising from the overlap of carbon p orbitals at right-angles to the plane of the molecule, and consequently electron density associated with this bond is located above and below the molecular plane. In general, transition metals are able to interact with this π electron density in a Lewis base–Lewis acid sense to form a stable unit: the π electron density on an alkene is donated into vacant orbitals on the transition metal.

6.2 Bonding of acyclic *polyhapto* ligands

The essential feature of the molecular-orbital description of the bonding between a metal and an alkene ligand was originally proposed by M. J. S. Dewar in 1951, and modified by J. Chatt and L. A. Duncanson in 1953: in addition to the donation of π electrons from the alkene to empty metal orbitals, there is also π-bonding between filled metal orbitals and the empty π* orbital of the alkene. We shall now consider the bonding in more depth, but it must be borne in mind that these bonding principles are not unique to alkenes. Indeed, similar or closely related descriptions have been discussed in relation to alkynes and to carbonyl complexes (Block 4, Section 3.3); they also apply to nitric oxide, dioxygen, dinitrogen, phosphine, arsine, isonitrile, 2,2′-bipyridyl, and related complexes. Included among these complexes are many catalysts of considerable industrial importance, and also some catalysts of biochemical importance such as haemoglobin. A feature of most of these complexes is that the transition metal occurs in a low oxidation state, often lower than the common oxidation states of the metals discussed earlier in the Course. Complexes of metals with low oxidation states are inherently thermodynamically unstable in the absence of π-acceptor ligands.

6.2.1 Bonding of *dihapto* (η^2) systems to metals

It is convenient to consider the bonding between a transition metal and an alkene as arising from two closely dependent components, namely a σ-bonding component and a π-bonding component, and to deal with these separately, at least initially. Let us consider the σ-bonding component in the Zeise's salt anion, $[PtCl_3(C_2H_4)]^-$.

SLC 7

Suitable bonding orbitals are formed when orbitals on the metal and alkene have similar energies, and when the signs of their *wavefunctions*, ψ, match in the bonding region. This, together with the geometry of the system, means that only certain metal and alkene orbitals can be involved in bonding.

The ethene π-molecular orbital is found to have a suitable energy for overlap with the platinum 6s, 6p and 5d orbitals.

☐ In separate diagrams, sketch the ethene π orbital and the platinum orbitals that have suitable symmetry for overlap with it. Assume that the platinum–ethene axis is the z axis.

■ Figure 25 shows that the platinum 6s, $6p_z$ and $5d_{z^2}$ orbitals have suitable symmetry for overlap with the ethene p orbital.

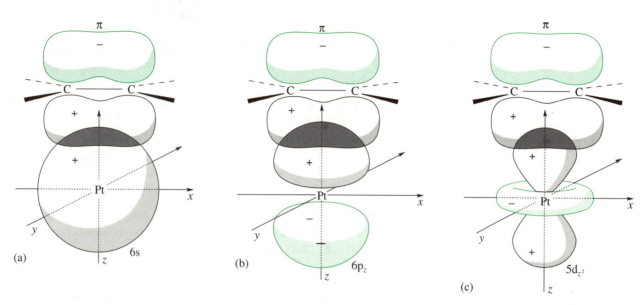

Figure 25 Diagram of the platinum orbitals with the correct symmetry to bond with the ethene π molecular orbital; (a), (b) and (c) lead to σ-bonding (no nodal planes).

The appropriate ethene orbitals — the σ-bonding and π-bonding molecular orbitals — must match the platinum orbitals in energy and must have a large overlap if strong bonding is to occur. The closer the match, the more stable will be the resulting platinum–ethene bonding molecular orbital.

The types of orbital overlap shown in Figure 25 are generally important for all transition metal *dihapto* alkene complexes; the contribution of each towards the bonding will depend on the relative energies of the s, p_z and d_{z^2} orbitals, which depend among other things on the position of the metal in the Periodic Table. A position towards the right of the Periodic Table tends to favour relatively less d-orbital involvement, because, as you know from earlier parts of the Course, in moving across a transition series, the d orbitals tend to become part of the core. If only metal–ligand σ-bonding were involved, most alkene complexes would not exist under normal conditions, because of the weak donor properties of alkenes. Additionally, for metals in low oxidation states, such a process would cause a build-up of electron density on the metal. This would tend to counteract the electron donation process, and so weaken the interaction, leading to instability of the metal–alkene bond. Obviously for transition-metal derivatives, this is not the situation that actually exists; hence a mechanism must occur for the dissipation of any charge build-up. The bonding system, known generally as **back-donation** or **back-bonding**, involves π-bonding between a filled metal orbital and an empty π* orbital of the alkene.

> Try to sketch this interaction before looking at Figure 26 (p. 42). Again, if we consider a simple metal–alkene unit such as in the Zeise's salt anion, $[PtCl_3(C_2H_4)]^-$, the platinum orbitals that have appropriate symmetry to overlap with the ethene π* orbital, are those shown in Figure 26, namely $6p_x$ and $5d_{xz}$.

Since the electrons associated with the platinum–ethene π bond were originally platinum electrons, and in the platinum–ethene complex are accommodated in a molecular orbital with some ethene character, metal electron density has effectively transferred from the platinum to the ethene.

The full σ + π-bonding picture may then be represented diagrammatically (Figure 27) as consisting of a σ donation of ethene π-electron density to the platinum, and a reverse process of π donation of metal electron density to the ethene π* molecular orbital. The two processes are not separable, but are **synergic**. Thus, the greater the σ donation, the greater the π back-donation from metal to alkene,

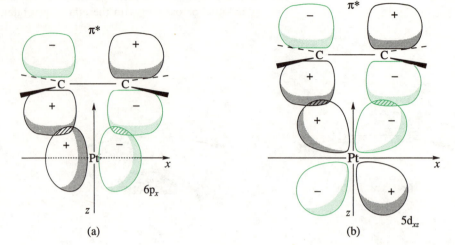

Figure 26 Bonding interactions between (a) the ethene π* orbital and the platinum $6p_x$ orbital, and (b) the ethene π* orbital and the platinum $5d_{xz}$ orbital for $[PtCl_3(C_2H_4)]^-$.

because of the greater tendency to cause a charge build-up at the metal. This synergic process is similar to that which operates for metal carbonyls, except that for carbonyls π-bonding occurs in two planes.

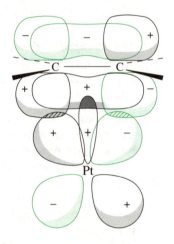

Figure 27 Diagrammatic representation of the σ (grey) and π (crosshatching) components of the platinum–ethene bonding interactions for $[PtCl_3(C_2H_4)]^-$.

A simple line diagram representation of the synergic process in the Zeise's salt anion is shown in Figure 28a.

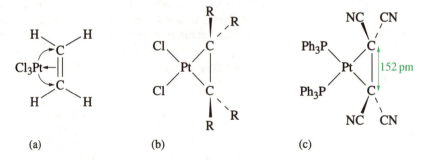

Figure 28 Two representations of the bonding in the platinum–ethene unit of the Zeise's salt anion and related complexes. In (a), bonding to the alkene is assumed to be primarily π-bonding, and in (b), to be primarily σ-bonding. (c) Structure of the complex $[Pt\{(NC)_2C=C(CN)_2\}(PPh_3)_2]$. (In general, note that C—C and C=C covalent bond lengths are 154 pm and 134 pm in alkanes and alkenes, respectively).

☐ How would you expect the synergic process to affect the bond order in ethene?

■ Electron density is *lowered* in the π-bonding orbital and *increased* in the π*-antibonding orbital, causing a lowering of the carbon–carbon bond order and a weakening of the bond. However, in the Zeise's salt anion this lengthening is minimal (C—C = 135 pm), so in this case the bond order is close to 2.

The consequences of the synergic process are illustrated by the lowering of the $v(C=C)$ stretching frequency by 50–150 cm^{-1} and by the increase in the carbon–carbon distances of coordinated alkenes compared with free alkenes. This latter feature is especially true for alkenes having substituents that are strongly electron

Figure 29 Representations of the σ and π orbital interactions for metal complexes of (a) alkenes and dioxygen, (b) carbon monoxide, nitric oxide and isonitriles, and (c) trifluorophosphine. Overlap in one plane only is shown. The σ-bonding interactions are shaded grey and the π-bonding interactions are indicated by crosshatching.

withdrawing, such as cyano and trifluoromethyl. For example, in the complex [Pt{$C_2(CN)_4$}(PPh$_3$)$_2$] (Figure 28c), where the alkene occupies two of the four coordination sites of the square-planar platinum, the alkene carbon–carbon distance of 152 pm is 18 pm longer than in free tetracyanoethene. This ligand may be regarded as having only weak σ-donor properties, but strong π-acceptor properties. The effect of extensive transfer of electron density into the alkene π* level such as occurs in this complex is to lengthen the alkene carbon–carbon distance to that characteristic of a carbon–carbon single bond. Thus, the overall effects of the synergic process are to balance the electron density in the π level with that in the π* level. If the two electron densities balance precisely, then the carbon–carbon multiple bonding has been destroyed. For such complexes, a three-membered-ring σ-bonding framework (Figure 28b and c) is an equally valid description. Note that the alkene carbon atoms for the tetracyanoethene complex are approximately sp^3 hybridised, whereas they were originally sp^2 hybridised. Thus, a feature of the bonding of alkenes to transition metals, in addition to an extension of the alkene carbon–carbon distance, is a lowering of the angle between the substituents at the carbon atoms, and their displacement away from the metal. The stronger the π-acceptor properties of the alkene, the longer the alkene carbon–carbon distance and the closer these alkene carbon atoms approximate to sp^3 hybridisation. The generality of the synergic effect is illustrated for alkene, dioxygen, carbon monoxide, nitric oxide, isonitrile and trifluorophosphine derivatives in Figure 29.

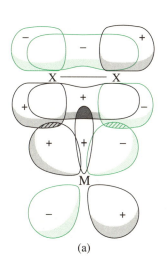
(a)

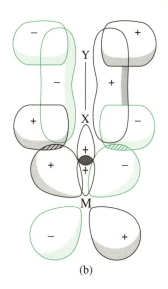
(b)

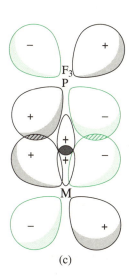
(c)

6.2.2 Bonding in other acyclic *polyhapto* derivatives

The bonding principles established for *dihapto* derivatives apply generally to other *polyhapto* derivatives, the essential features being that the molecular orbitals of the organic molecule or group are matched with those metal orbitals of appropriate energy and symmetry about the metal–ligand axis. The process is slightly more complicated than for an η^2-alkene only because additional ligand orbitals are involved. The considerations are otherwise the same.

SAQ 7 Draw diagrams to represent the interaction of the p_z orbitals of the skeletal carbon atoms for the allyl group, CH_2CHCH_2. The z axis is at right-angles to the plane of the three carbon atoms. As there are three atomic orbitals, three molecular orbitals should be obtained, one bonding, one non-bonding and one antibonding. Then assign metal atomic orbitals with the correct symmetry to overlap with each of the molecular orbitals of the allyl group, assuming that the metal is positioned above the plane of the allyl group.

☐ Now try drawing diagrams to represent the interaction of the p_z orbitals of the carbon atoms of buta-1,3-diene, to give four sets of atomic orbital combinations, given that two correspond to bonding and two to antibonding levels. (Again, assume that the z axis is perpendicular to the plane of the carbon atoms.) Then assign metal atomic orbitals with the correct symmetry to overlap with each of the molecular orbitals of the butadiene, assuming that the metal is positioned above the plane of the diene.

- The orbitals of buta-1,3-diene and the appropriate corresponding metal orbitals in order of increasing energy are shown in Figure 30. Orbitals with the same symmetry are in the same rows.

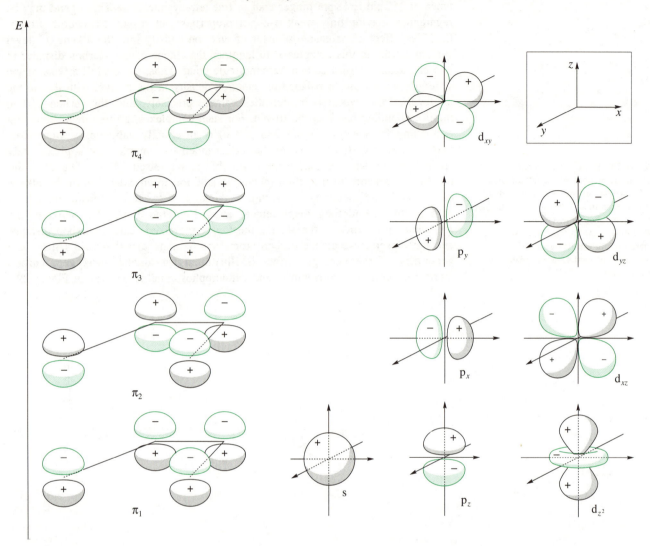

Figure 30 Combinations of carbon p_z orbitals of buta-1,3-diene with metal atomic orbitals having appropriate symmetry for bonding.

It is instructive to consider the possible implications of these various bonding interactions between a metal and buta-1,3-diene. Three extreme situations can be envisaged, represented by Figure 31a, b and c. In each case, the C_4H_6 group acts formally as a four-electron η^4-ligand, but, if the bonding interactions are of the same energy, then Figure 31a implies that π_1 makes the major contribution, Figure 31b implies that π_2 makes the major contribution and Figure 31c that π_3 makes the major contribution to the bonding. Each of the above representations implies particular carbon–carbon bond lengths. For example, for Figure 31a the three lengths should be approximately equal, for Figure 31b the two peripheral carbon–carbon bonds should be shorter than the central bond, and the opposite situation should occur for Figure 31c. These arguments apply equally to all 1,3-diene complexes. Structural data are of obvious importance here, and data for some suitable complexes are given in Figure 32. It is immediately apparent that the carbon–carbon distances differ markedly from complex to complex. The distances in Figure 32a and c are clearly compatible with the bonding descriptions in Figure 31a and c, respectively, but the bond lengths in Figure 32b only approximately relate to the π,π interactions of Figure 31b.

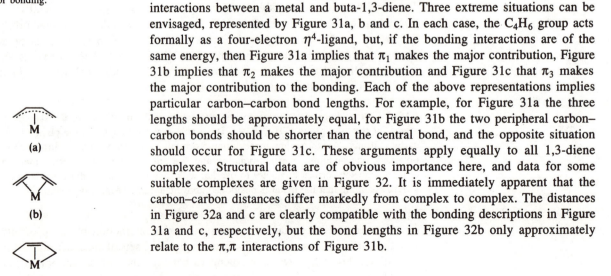

Figure 31 Bonding interactions between a transition metal and buta-1,3-diene: (a) delocalised π interaction; (b) π,π interactions; (c) σ,σ,π interactions.

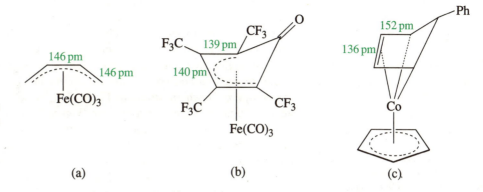

Figure 32 Structures of some 1,3-diene complexes: (a) tricarbonylbutadienyliron; (b) tricarbonyltetratrifluoromethyl-cyclopentadienoneiron; (c) η^5-cyclopentadienyl-1-*exo*-phenyl-cyclopentadienecobalt.

☐ Consider now the implications of the sequence of π energy levels for buta-1,3-diene, namely $\pi_1 < \pi_2 < \pi_3 < \pi_4$. Assign four electrons to these levels in the ground state, and then deduce the configuration of the first excited state. What changes in geometry would you expect for the buta-1,3-diene molecule in going from the ground state to the first excited state?

■

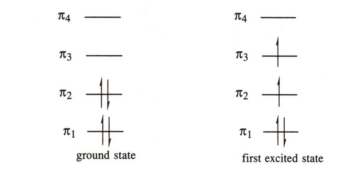

In the ground state of the free molecule, π_1 and π_2 would each accommodate two electrons, those of the latter orbital being particularly responsible for the short/long/short sequence of carbon–carbon bond lengths. In the first excited state, an electron is transferred from π_2 to π_3, the orbital concerned predominantly with π-bonding between the central carbon atoms. Thus, conversion to a structure with equal-length carbon–carbon bonds is predicted.

The bonding of butadiene to a metal is dependent on the nature of the metal. The depletion of electron density in π_2 arises from electron donation to a metal, and occupation of π_3 results from electron donation from the metal. A trend from equal-length carbon–carbon distances to a long/short/long sequence is expected to reflect the extent of the σ-donation and π back-donation processes. Look back to Figure 31, in which the bond lengths of a diverse series of 1,3-diene complexes are given. From the above discussion we see that the equal bond lengths observed in the iron complexes are not simply due to a π_1 combination of orbitals, but are also due to occupation of π_2 and π_3. Although the organo groups in the complexes differ slightly in terms of their substituent groups, it is apparent that the other ligands in 1,3-diene complexes have a profound effect on the diene–metal bonding.

☐ Try to rationalise the trends in 1,3-diene bond lengths in terms of the effect of other ligands on the metal.

■ When alkyl groups are present, back-bonding to the diene occurs more strongly than in carbonyl derivatives. This is to be expected, since carbonyl groups are more electron-withdrawing than alkyl groups and so compete more strongly for the available metal π-electron density. The overall effect is that in complexes containing alkyl groups, the electron density in π_3 is greater than that for related carbonyl complexes, and that the long/short/long sequence of skeletal carbon–carbon distances is more pronounced.

Figure 33 Examples of alkyne complexes:
(a) [Mn(η^5-C$_5$H$_5$)(CO)$_2$(PhC≡CPh)], a monoalkyne, mononuclear complex;
(b) [{PtCl$_2$(Bu^{t}C≡CBu^t)}$_2$], a *bis*-alkyne, binuclear complex;
(c) [Fe$_3$(CO)$_9$(PhC≡CPh)], a monoalkyne, trinuclear complex.

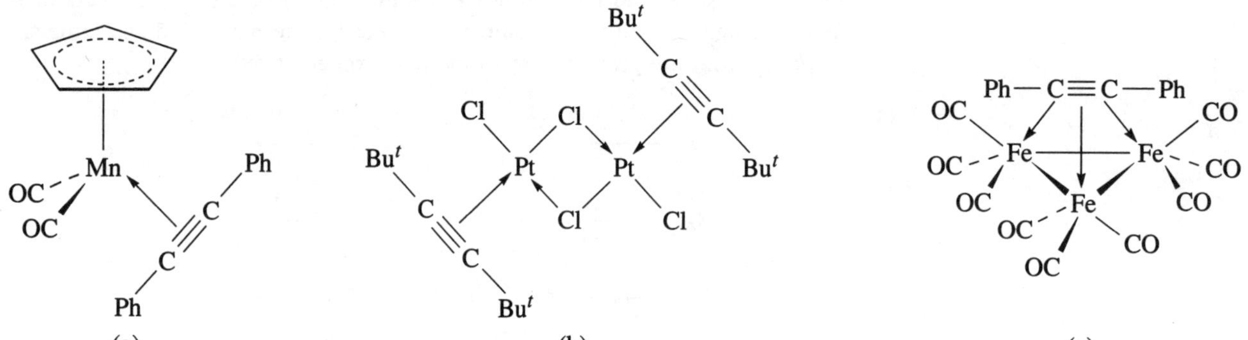

Alkynes also have a π system that could be involved in η^2-bonding like that in metal–alkene complexes. This is borne out by the formation of stable alkyne complexes analogous to alkene complexes, such as [Mn(η^5-C$_5$H$_5$)(CO)$_2$(PhC≡CPh)] and [{PtCl$_2$(Bu^{t}C≡CBu^t)}$_2$] (Figure 33a and b), but as alkynes have two π bonds at right-angles to each other you might also expect that two metals could be bound to each carbon atom. This possibility has been realised; see, for example, the complex shown in structure **IV** of SAQ 8 below. Alkynes are, in fact, most versatile ligands, such that trinuclear (Figure 33c) and tetranuclear complexes are also known. There is a variety of bonding in alkyne complexes, and the triple bonds in Figure 33 should not be taken as indicating no modification in bonding pattern from the parent alkyne. The stability of some of the polynuclear complexes, such as the one in Figure 33c, is probably enhanced by strong metal–metal bonding within the clusters of metal atoms. You will learn how to predict the shapes of these organometallic clusters by simple application of electron counting rules later in this Section.

6.3 The *polyhapto* classification of organic ligands

The alternative ways of describing the bonding of alkenes to transition metals, particularly when strongly electron-withdrawing substituents are present, make classification according to bond type rather artificial. Traditionally, alkene complexes and derivatives of other molecules or groups containing π systems, which interact strongly with metals, have been classified as π complexes, but it should be apparent that the bonding of some alkenes is best considered in terms of σ-bonding between the metal and the alkene carbon atoms in a three-membered ring, with no significant degree of π-bonding. Consequently, you will now have a greater understanding of why we generally use the *hapto* terminology (or *hapticity*) to indicate the number of atoms of an organo group that are considered to be within bonding distance of the metal. This then avoids the need to imply a particular type of bonding. Ethene complexes, for example, will be referred to as *dihapto* ethene complexes, or η^2-C$_2$H$_4$ in formulae.

☐ [Fe(CO)$_5$] reacts with buta-1,3-diene to form complexes [Fe(C$_4$H$_6$)(CO)$_4$] and [Fe(C$_4$H$_6$)(CO)$_3$]. Assuming that the number of electrons available to the iron atom is unchanged, suggest an appropriate prefix to C$_4$H$_6$ to indicate what you would expect the attachment of the diene to be in each of the complexes.

■ Butadiene replaces one carbon monoxide group in [Fe(CO)$_5$] to form [Fe(C$_4$H$_6$)(CO)$_4$], but two carbon monoxides to form [Fe(C$_4$H$_6$)(CO)$_3$]. Thus, C$_4$H$_6$ acts as a two-electron donor in [Fe(C$_4$H$_6$)(CO)$_4$] and a four-electron donor in [(Fe(C$_4$H$_6$)(CO)$_3$]. It follows from the previous Section that if C$_4$H$_6$ uses just two electrons in bonding to the metal, then it acts as a monoalkene and is *dihapto* or η^2-C$_4$H$_6$. But in [Fe(C$_4$H$_6$)(CO)$_3$], the butadiene is a four-electron or dialkene-type ligand, and hence a *tetrahapto* or η^4-ligand. The structure of this latter complex was shown as Figure 32a.

SAQ 8 For the organo ligands in the following complexes **I–V**, (a) give the *hapticity*, (b) specify them as σ- or π-bound ligands, and (c) say whether they are μ-bonded or non-bridged ligands. Your attention should be directed only at the organo ligand.

$(OC)_3Fe$ —[cyclooctatetraene]— $Fe(CO)_3$

I

[Ir complex with R_3P, CO, and $C(CN)_2-C(CN)_2$ chelate]

II

[cyclopentadienyl with $Fe(CO)_3$ above and $Fe(CO)_3$ below]

III

$Ph-C\equiv C-Ph$ bridging $(OC)_3Co-Co(CO)_3$

IV

$OC-W(CO)_3=C(Ph)(OMe)$ carbene

V

6.4 'Bucky balls'

Until 1986, carbon was thought to commonly exist in the two allotropic forms graphite and diamond. Probably the most exciting facet of organometallic chemistry at the present time is the discovery of **Buckminsterfullerene** or bucky balls[†]. This new elemental form of carbon was predicted from astronomical spectra, and can now be synthesised by arc-welding graphite rods under a partial vacuum containing helium. This process generates a number of high-mass clusters of carbon atoms, detected initially by mass spectrometry, typically with clusters in the range C_{40}–C_{128}. However, certain clusters — C_{60}, C_{70}, C_{84} — commonly known as *fullerenes*, proved to be particularly abundant. Sophisticated chromatographic techniques have enabled chemically significant amounts of the individual clusters to be isolated. The C_{60} cluster has achieved the greatest attention so far, since it is the easiest to make and isolate. Difficulty was found in assigning the molecular structure of C_{60} using X-ray crystallography, because of rapid spinning of the C_{60} cluster even at low temperature. The key step to defining the geometry unambiguously involved the formation of organometallic complexes of C_{60}, thus introducing molecular 'hooks' to the surface to minimise rotation. This was achieved by synthesising $[OsO_4C_{60}(Bu^tC_5H_4N)_2]$ and $[Pt(\eta^2-C_{60})(PR_3)_2]$ by reaction of C_{60} with $[OsO_4]/Bu^tC_5H_4N$ and $[Pt(PR_3)_3]$, respectively.

The C_{60} cluster was found to be shaped like a soccer ball — a truncated icosahedron (Figure 34a). The carbon atoms are all three coordinate and the surface can be seen to consist of six-membered rings joined at the edges by five-membered rings. The structure can be considered to have thirty double bonds, delocalised across the framework. The ^{13}C n.m.r. spectra of C_{60} shows only one resonance.

Multiply substituted C_{60} has now been made in the form of the platinum derivatives $[\{Pt(PR_3)_2\}_x(\eta^2-C_{60})]$ ($x = 1, 2, 3, 4, 5, 6$). These contain η^2-bound C_{60} and are analogous to the Zeise's salt anion, $[PtCl_3(\eta^2-C_2H_4)]^+$, in the mode and manner of bonding (Section 6.2); see Figure 34b.

Another facet of C_x chemistry ($x = 60, 70$, etc.) is that by using special techniques the C_x species can be generated in the gaseous phase with encapsulated metal ions (for example La^{3+}) inside the fullerenes. It is also possible to form organometallic gas-phase adducts MC_{60} with metals like Fe, Ni and Co on the outside of the fullerene.

[†] This name was derived from Buckminster Fuller, an American architect famous for his geodesic sphere designs.

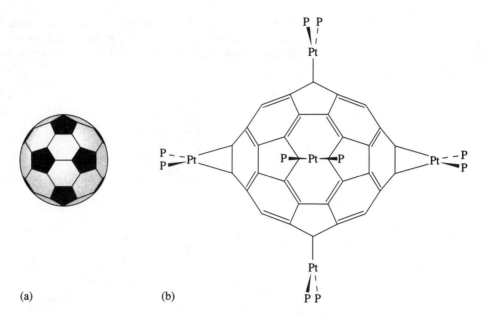

Figure 34 (a) A soccer ball; (b) [{Pt(PR$_3$)$_2$}$_x$(η^2-C$_{60}$)]. Note the similarity between these shapes.

The organometallic chemistry of C$_{60}$ and the higher fullerenes is currently (1994) a rapidly growing area, especially as purification and synthesis (albeit still at very high energy) techniques have now been elucidated. Buckminster fullerenes look to have potential as lubricants and superconductors, although some of the claims about their properties are yet to be substantiated.

6.5 The eighteen-electron rule

Earlier in the Course, we outlined the eighteen-electron rule; this rule can be usefully applied to organometallic complexes. Implicit in this rule is the observation that, for many complexes, the number of valence-shell electrons available to the metal is the same as that of the next noble gas. The eighteen electrons will be accommodated in nine valence-shell orbitals derived from the metal atom's five $(n-1)$d orbitals, the three np orbitals and the one ns orbital (remember that n is the principal quantum number of the metal). These electrons will comprise those donated by the ligands, together with the m electrons associated with the d^m configuration of the metal. Consequently, if it is known that a particular class of complex follows the rule (as many do), then the rule can be used to predict the formulation and coordination number of the metal for individual complexes. As, however, it is often impossible to assign meaningful oxidation states to metals and charges to ligands, particularly when extensive molecular orbitals are involved in their bonding, we shall adopt the convention in this Block that the metal is in oxidation state zero for the purpose of electron counting, even when it is actually in some other state, and treat each ligand as a neutral molecule or as a radical. (When the eighteen-electron rule was discussed in Block 4, the metal was regarded as having a formal—usually non-zero—oxidation state.)

We count up the electrons using the following rules. For metal carbonyls, a terminal group contributes two electrons to the metal, and a bridging carbonyl group a total of two electrons to all the metal atoms involved. A metal–metal single bond donates one electron to each metal. A terminal halogen donates one electron to the metal; bridging halogens contribute a total of three electrons to the two metal atoms. Simple alkenes use two electrons for bonding to metal atoms, butadiene up to four electrons and benzene up to six electrons. An alkyl group provides one electron, an allyl group, −CH$_2$CH=CH$_2$ (which may be either *monohapto* or *trihapto*), provides one or three electrons, and a cyclopentadienyl group, C$_5$H$_5$ (which may be *monohapto*, *trihapto*, or *pentahapto*), provides one, three or five electrons. In general, then, the *hapto* number and the number of electrons associated with the organic fragment are the same. The rules are summarised in Table 2.

Table 2 Number of electrons donated from various ligands in different coordination modes for application of the eighteen-electron rule

Ligands	η^1	μ_2	Other modes
CO	2	2	$\mu_3 = 2$
Cl, Br, I	1	3	
alkyl group R = Me, Et, Pr, Ph	1	1	
acyl group	1	1	
cyclopentadienyl	1	1	$\eta^3 = 3, \eta^5 = 5$
PR$_3$	2	2	
ethene	—	—	$\eta^2 = 2$

We can illustrate this new approach for [MnCl(CO)$_5$]; the manganese atom has ten electrons available from five CO groups, one electron from chlorine and seven from manganese (d^{5}s^2 configuration), giving a total of eighteen electrons.

☐ Why does a terminal Cl donate one electron, and a CO and PR$_3$ group donate two electrons for electron-counting purposes when the metal is considered to be in oxidation state zero ?

■ Chlorine (s^{2}p^5) donates an electron to a metal from its half-filled orbital when terminally bound. CO and PR$_3$ donate electrons via lone pairs, and are thus two-electron donors. When chlorine acts as a bridging ligand, it uses one lone pair as well its singly occupied orbital, so it is then a three-electron donor.

☐ Suggest the formulae of uncharged complexes that obey the eighteen-electron rule containing:
(a) carbon monoxide and/or benzene bonded to chromium;
(b) cycloocta-1,3-diene and nickel;
(c) CF$_3$ and/or CO groups bonded to iron.

■ (a) Cr (4s^{1}3d^5) provides six electrons and benzene provides six electrons, so [Cr(η^6-C$_6$H$_6$)$_2$] is an eighteen-electron system. Alternatively, C$_6$H$_6$ is equivalent to three CO ligands in terms of available electrons, so [Cr(η^6-C$_6$H$_6$)(CO)$_3$] and [Cr(CO)$_6$] are also eighteen-electron systems: these complexes are well known and stable. Other alternatives, such as [Cr(η^2-C$_6$H$_6$)(CO)$_5$], [Cr(η^4-C$_6$H$_6$)(CO)$_4$], [Cr(η^2-C$_6$H$_6$)(η^6-C$_6$H$_6$)(CO)$_2$], [Cr(η^4-C$_6$H$_6$)$_2$(CO)$_2$], and [Cr(η^4-C$_6$H$_6$)(η^6-C$_6$H$_6$)(CO)], also obey the eighteen-electron rule and are theoretically correct answers, but in practice they are unstable towards the loss of CO and formation of [Cr(η^6-C$_6$H$_6$)(CO)$_3$] and [Cr(CO)$_6$].

(b) *Tetrahapto* dienes provide four (2 + 2) electrons and nickel (4s^{2}3d^8) provides ten electrons. Thus, the eighteen-electron complex is [Ni(C$_8$H$_{12}$)$_2$], as long as the cyclo-octa-1,3-diene ligands are both *tetrahapto*.

(c) Iron (4s^{2}3d^6) provides eight electrons, so to achieve a total of eighteen, ten additional electrons are required. These could be provided by five × two-electron donors (CO) to form [Fe(CO)$_5$], or two × one-electron donors (CF$_3$) and four × two-electron donors (CO) to form [Fe(CO)$_4$(CF$_3$)$_2$].

The eighteen-electron rule is followed by the vast majority of organometallic complexes of the d-block elements in which they are in a low oxidation state, although exceptions do occur towards the two extremes of the d-block in the Periodic Table. Metals with d^8 configurations (Pt^{II}, Pd^{II}, Ir^{I}, Au^{III}) have a particularly strong tendency to form sixteen-electron square-planar complexes.

SAQ 9 Suggest how eighteen bonding electrons may be achieved for each metal in the following compounds, for which the empirical formulae only are given:
(a) ReCl(CO)$_4$; (b) FeI(CO)$_4$; (c) MnCl(CO)$_6$.

6.6 Isolobal analogy

We have seen in the previous Section that the eighteen-electron rule can be used as a guide to the stability and geometry of transition-metal complexes. In the next Section we shall see how the Wade–Mingos rules can be used to predict the shape of many small clusters of metal atoms. But first we shall consider how electron counting can be used to predict which molecular fragments can bond together — the isolobal analogy. These two principles have helped to systematise organo-metallic chemistry in recent years by pulling together what was once a huge series of seemingly unrelated reactions and compounds into a more cohesive whole.

In 1981 R. Hoffman and K. Fukui were awarded the Nobel Prize for Chemistry in part for developing a concept called the **isolobal analogy**, which, although intuitively simple, has great predictive power. The essence of their theory is that isolobal molecular fragments can bond together to form compounds.

> The two fragments under comparison are considered to be isolobal if the number, symmetry properties, approximate energy and shape of their valence orbitals, as well as the number of electrons occupying them, are similar.

'Valence orbitals' are defined as those that are available for bonding interactions. Here, 'analogy' means the comparison of two or more molecular fragments.

The isolobal analogy is used to compare two or more molecular fragments with regard to the shape and electronic occupation of their valence orbitals. This comparison allows us to predict what fragments may combine together to form a compound by matching isolobal fragments. For our purposes, the energy and shape of the valence orbitals are less important in determining whether fragments are isolobal than is the electron occupancy of these orbitals.

In simplified terms, for fragments to be isolobal there must be the same number of unoccupied and half-occupied orbitals in the valence orbitals of the two fragments under consideration. Furthermore the **electron-count deficiency** in the valence orbitals of each fragment must be the same. If a fragment is electron deficient, it will be able to share electrons with a similar electron-deficient fragment in order to reach a stable electronic configuration, thus forming bonds. For a fragment containing a main-Group central atom, the electron-count deficiency is calculated as $8 - (x + y)$, where x = the number of valence electrons associated with the central atom, and y = the number of electrons provided by the attached group of ligands.

So, for the $\cdot CH_3$ radical, x (the number of carbon valence electrons) = 4, and $y = 3$, since each H atom donates one electron. This gives the electron-count deficiency as $8 - (4 + 3) = 1$. For fragments containing a transition metal it is assumed that the metal obeys the eighteen-electron rule, so the electron-count deficiency is given by $18 - (x + y)$.

☐ What is the electron-count deficiency for $Mn(CO)_5$?

■ As manganese is a transition metal, we use $18 - (x + y)$. Mn has seven valence electrons ($3d^5 4s^2$), so $x = 7$. Each carbonyl ligand donates two electrons, so $y = 10$, giving $18 - (7 + 10) = 1$.

We can see that $\cdot CH_3$ and $Mn(CO)_5$ both have a deficiency of one electron. But to be isolobal they must also have the *same number* of *half-filled* valence orbitals. For main-Group species the central atom can have only four orbitals (ns, np_x, np_y and np_z). In addition, transition metals have five d orbitals of appropriate energy, giving them a total of nine valence orbitals (ns, three np and five $(n-1)d$) to be filled. Thus, as $\cdot CH_3$ has seven electrons and four valence orbitals (each orbital holds two electrons), there is one half-filled orbital.

☐ Now work out how many half-filled valence orbitals $Mn(CO)_5$ has.

■ Mn(CO)$_5$ has seventeen electrons (seven from Mn and ten from the five CO ligands). As there are nine manganese orbitals, one is half-filled.

·CH$_3$ and Mn(CO)$_5$ therefore have the same electron-count deficiency (one) *and* the same number of half-filled orbitals (one). They are said to be *isolobal* and this is represented symbolically as follows:

$$\text{Mn(CO)}_5 \longleftrightarrow \cdot\text{CH}_3$$

In other words, ⟵⟶ is the symbol used to denote isolobal fragments.

The theory predicts that ·CH$_3$ and Mn(CO)$_5$ could pair up their half-filled orbitals to form the complex [Mn(CO)$_5$(CH$_3$)].

For an ML$_6$ complex that obeys the eighteen-electron rule, if one two-electron ligand is removed to make an ML$_5$ fragment (such as loss of CO from [Mn(CO)$_5$(CH$_3$)]), then there are sixteen electrons left and one vacant orbital. Figure 35b (p.52) shows simplified molecular-orbital diagrams for such a system.

In Figure 36 we see this then continued for ML$_5$, ML$_4$ and ML$_3$ fragments. Examples are shown for Mn(CO)$_5$ (d^7), Fe(CO)$_4$ (d^8) and Co(CO)$_3$ (d^9).

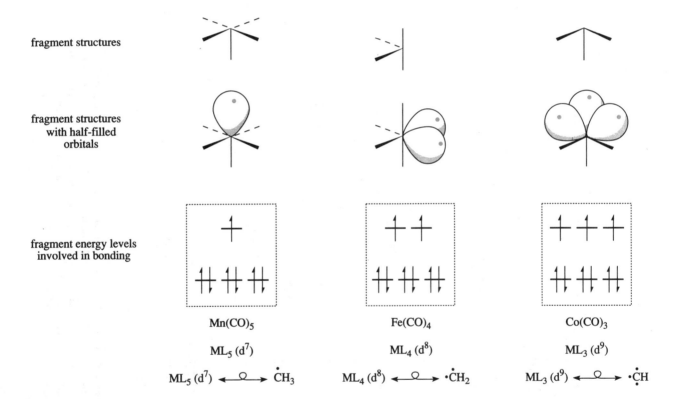

Figure 36 Fragment structures, half-filled σ-bonding orbitals and energy levels involved in bonding for ML$_5$ (d^7), ML$_4$ (d^8), ML$_3$ (d^9). For ML$_5$ (d^7) one σ-bonding orbital contains one electron. For ML$_4$ (d^8), two σ-bonding orbitals each contain an unpaired electron. For ML$_3$ (d^9), three σ-bonding orbitals each contain one unpaired electron.

☐ How many σ-bonding orbitals would the ML$_4$ and ML$_3$ systems be expected to have?

■ Four and three, respectively.

☐ What is the total electron count for Mn(CO)$_5$, Fe(CO)$_4$ and Co(CO)$_3$?

■ Mn(CO)$_5$: Mn is d^7, giving an electron count of 7 + 10 = 17;
Fe(CO)$_4$: Fe is d^8, giving an electron count of 8 + 8 = 16;
Co(CO)$_3$: Co is d^9, giving an electron count of 9 + 6 = 15.

Feeding the electrons into the molecular orbital diagrams in Figure 36 for each fragment, one singly occupied orbital is generated for manganese, two for iron and three for cobalt. (Hund's rule says that electrons will singly occupy degenerate orbitals in preference to spin pairing.)

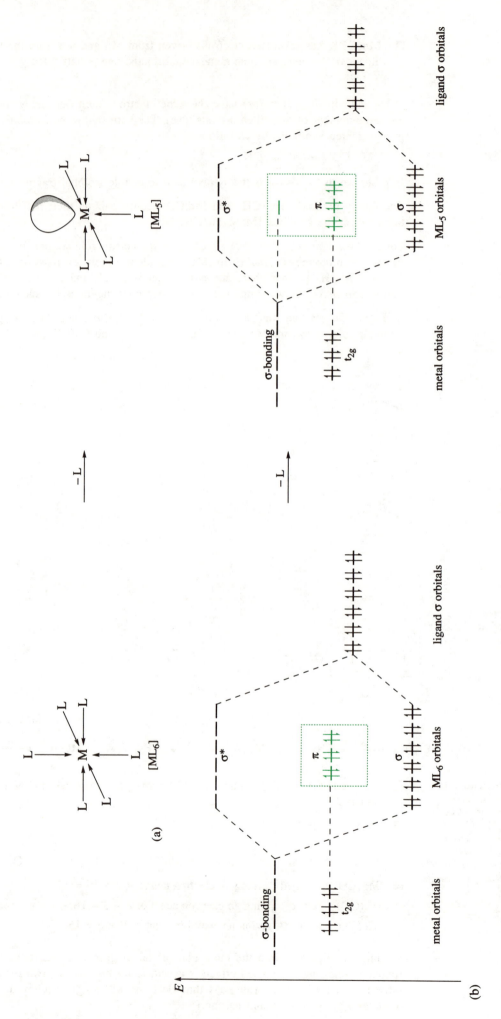

Figure 35 (a) ML$_6$ and ML$_5$ fragment structures; removal of one ligand, L, from an eighteen-electron complex ML$_6$ creates one new valence orbital. (b) Simplified molecular orbital energy-level schemes for an ML$_6$ complex (left) and an ML$_5$ complex (right). The metal orbitals on the left of each diagram consist of three t_{2g} orbitals (d_{xy}, d_{yz}, d_{xz}) and six σ-bonding orbitals (s, p$_x$, p$_y$, p$_z$, d$_{x^2-y^2}$, d$_{z^2}$), grouped together for simplicity. Six ligand orbitals are represented on the right of the ML$_6$ diagram; for ML$_5$, one ligand has been removed and hence one of the six metal σ-bonding orbitals remains unpaired. Orbitals in the broken boxes are the ones involved in bonding.

□ Predict whether the radical ·ĊR is isolobal with the d^9 ML_3 fragment $Rh(CO)_3$.

■ The electron-count deficiency for $Rh(CO)_3$ is $18 - (9 + 6) = 3$, and for ·ĊR is $8 - (4 + 1) = 3$. $Rh(CO)_3$ has fifteen valence electrons $(9 + (3 \times 2))$, and from Figure 36 this gives three half-occupied orbitals. ·ĊR has five valence electrons, four from carbon and one from R. Main-Group elements have eight electrons in a closed valence shell, so as ·ĊR has five valence electrons this leaves three half-filled orbitals. Thus, $Rh(CO)_3$ is isolobal with ·ĊR:

$$Rh(CO)_3 \longleftrightarrow ·ĊR$$

From this analogy we could imagine forming a molecule such as $R-C\equiv Rh(CO)_3$ by combining the two fragments. Then the carbon would effectively have eight electrons and rhodium would have eighteen electrons in their outer valence shells.

The beauty of the isolobal approach is that it enables us to predict which fragments can combine together to form a molecule; Figure 37 illustrates some possibilities.

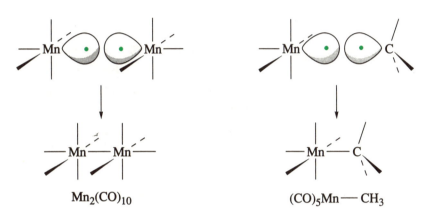

Figure 37 Isolobal comparison for ·CH$_3$ and ·Mn(CO)$_5$, showing possible combinations.

□ Which of the following fragments are isolobal: $Fe(CO)_4$, $Fe(CO)cp^-$, $Rh(CO)cp$, ·ĊR$_2$, As? Use the isolobal analogy to decide on possible dimeric compounds formed from two of the fragments. (Note that cp$^-$ stands for cyclopentadienyl and acts as three ligands, 3L.)

■ $Fe(CO)_4 \equiv ML_4$ (d^8); the electron-count deficiency is given by $18 - (8 + 8) = 2$. There are sixteen valence electrons, so there are *two* half-filled orbitals.

$Fe(CO)cp^- \equiv ML_4$ (d^8); the electron-count deficiency is given by $18 - (8 + 2 + 5 + 1) = 2$. There are sixteen valence electrons, so there are *two* half-filled orbitals. (Note that the negative charge on the $Fe(CO)cp^-$ is included in the electron counting.)

$Rh(CO)cp \equiv ML_4$ (d^9); the electron-count deficiency is given by $18 - (9 + 2 + 5) = 2$. There are sixteen valence electrons, so there are *two* half-filled orbitals.

·ĊR$_2$; the electron-count deficiency is $8 - (4 + 2) = 2$. There are six valence electrons, so there are *two* half-filled orbitals.

As; the electron-count deficiency is $8 - (5) = 3$. There are five valence electrons, so there are *three* half-filled orbitals.

Thus,

$$Fe(CO)_4 \longleftrightarrow Fe(CO)cp^- \longleftrightarrow Rhcp(CO) \longleftrightarrow ·ĊR_2 \not\longleftrightarrow As$$

We know that alkenes ($R_2C=CR_2$) exist. Therefore, as ·ĊR$_2$ is isolobal with $Fe(CO)_4$, [$(CO)_4Fe=CR_2$] (a Fischer carbene) is predicted to exist by the isolobal analogy. Furthermore, a whole series of combinations is theoretically possible, such as [$(CO)_4Fe=Fe(CO)_4$], [$(CO)cpFe=Fe(CO)cp]^{2-}$, [$(CO)_4Fe=Fe(CO)cp]^-$, [$(CO)cpRh=CR_2$], and so on.

The isolobal analogy has now gained wide acceptance in the chemical community. It gives us predictive power in deciding which fragments could form bonds by simple orbital correlation. However, this approach does predict molecules that cannot be synthesised because of a lack of suitable reagents, or due to chemical incompatibility leading to highly kinetically unstable molecules.

In summary:

ML_5 (d^7) $d^7 + 10 = 17$ electrons, one half-filled orbital;

ML_4 (d^8) $d^8 + 8 = 16$ electrons, two half-filled orbitals;

ML_3 (d^9) $d^9 + 6 = 15$ electrons, three half-filled orbitals.

ML_5 (d^7) ⟵⟶ ·CR_3; ML_4 (d^8) ⟵⟶ ·$\ddot{C}R_2$; ML_3 (d^9) ⟵⟶ ·$\ddot{C}R$.

SAQ 10 State which of the following fragments are isolobal:

(a) $Mn(CO)_5$, ·CH_3, As;

(b) ·$\ddot{C}H$, P, BH^-, S^+, Se;

(c) ·$\ddot{C}H$, As, $Co(CO)_3$, Sn^-;

(d) $Re(CO)_4^-$, $Os(CO)_4$, ·$\dot{C}H_2$;

(e) $Ir(PR_3)_4$, ·CH_3, SR_2, PR_3, AsR_3;

(f) $Ni(CO)_2$, $Os(CO)_3$, BH;

(g) $[Ni_5(CO)_{10}]^{2-}$, $[B_5H_5]^{2-}$;

(h) $[Os_5(CO)_{15}]^{2-}$, $[B_5H_5]^{2-}$.

(*Hint* In parts (g) and (h), consider the parent fragment — for example $Ni(CO)_2$ for $[Ni_5(CO)_{10}]^{2-}$, etc. If the parent fragments are isolobal, then so will any clusters formed from them. Thus, if A ⟵⟶ B then $(A)_n$ ⟵⟶ $(B)_n$).

6.7 Wade–Mingos rules

In this Section you will learn how electron counting can allow the prediction of the geometry of both organometallic and main-Group compounds. The Wade–Mingos theory, which was developed by Professor K. Wade and Professor M. Mingos, is quite complicated. You will not be expected to answer detailed examination questions on this topic, but it is important that you attempt all of SAQ 11, since the principles of the Wade–Mingos theory are best learnt by example.

The **Wade–Mingos rules** enable us to predict the geometry of organometallic clusters of atoms, which in turn allows us to rationalise the bonding in quite complicated molecules.

Unlike the eighteen-electron rule, which uses as its basis the localised bond (two-electron, two-centre), the Wade–Mingos theory treats the molecule as a whole and considers the number of electron pairs and vertices. It has a basis in classical geometry. Although a detailed understanding is beyond the scope of this Course, the application of the rules will enable us to determine the structure of many organometallic complexes, and clusters.

First we need to be able to recognise the simple polyhedra drawn in Figure 38. The Wade–Mingos theory proposes that clusters of transition-metal and main-Group elements will be based on these polyhedra. The vertices of each polyhedron are occupied by a metal atom or main-Group fragment, and the geometry is dependent on the total number of electrons present in the cluster.

For a cluster containing n metal atoms or fragments, the Wade–Mingos rules tell us that if there are $n + 1$ skeletal electron pairs then the cluster will be based on a polyhedron that has n vertices, and that each vertex will be occupied by a metal atom. Such a cluster is said to be *closo* because the polyhedron is complete; that is, each vertex is occupied.

The rules go on to say that if there are *more* than $n + 1$ skeletal electron pairs, then the shape of the cluster will be based on a polyhedron with more than n

3 vertices
(4 electron pairs)
triangular

4 vertices
(5 electron pairs)
pyramidal

5 vertices
(6 electron pairs)
trigonal bipyramidal

6 vertices
(7 electron pairs)
octahedral

6 vertices
(7 electron pairs)
trigonal prismatic

7 vertices
(8 electron pairs)
pentagonal
bipyramidal

8 vertices
(9 electron pairs)
cunane

8 vertices
(9 electron pairs)
cubic

8 vertices
(9 electron pairs)
twisted cube

Figure 38 Polyhedra with electron counts for **closo** structures — that is, no vacancies. (Note that not all possible shapes for each number of vertices are shown).

vertices, but that some of the vertices will be unoccupied; that is, the structure may be regarded as 'opening out'. We shall enlarge on this later.

To use this treatment we start off by subdividing a cluster into fragments, which we shall now call **structural units**. Each structural unit contains one metal atom. When the structural units are assembled, they will together constitute the whole structure. Taking the $Os_5(CO)_{16}$ cluster as an example, there are five osmium atoms and sixteen carbon monoxide ligands; the osmium atoms constitute the vertices of the polyhedron. We then distribute the carbon monoxide ligands as evenly as possible around the osmium atoms. This results in four $Os(CO)_3$ structural units and one $Os(CO)_4$ structural unit.

Once we have decided on our structural units, we need to evaluate the number of skeletal electrons present in each structural unit and calculate an *overall* electron count for the cluster. We then compare the number of structural units with the number of electron pairs, which enables us to predict the cluster geometry according to the correlations in Table 3.

Table 3 Correlation between numbers of electron pairs and molecular shape, using the Wade–Mingos rules

	Number of electron pairs	Number of structural units, n	Structural type classification	Shape
$n + 1$	4	3	*closo*	trigonal
	5	4	*closo*	pyramidal
	6	5	*closo*	trigonal bipyramidal
	7	6	*closo*	octahedral, or trigonal prism
	8	7	*closo*	pentagonal bipyramidal
	9	8	*closo*	cunane or cubic or twisted cube
$n + 2$	6	4	*nido*	trigonal bipyramidal, one vertex missing
	7	5	*nido*	octahedral, one vertex missing
	8	6	*nido*	pentagonal bipyramidal, one vertex missing
	9	7	*nido*	cubic, one vertex missing
$n + 3$	6	3	*arachno*	trigonal bipyramidal, two vertices missing
	7	4	*arachno*	octahedral, two vertices missing
	8	5	*arachno*	pentagonal bipyramidal, two vertices missing
	9	6	*arachno*	cubic, two vertices missing
$n + 4$	8	4	*hypo*	pentagonal bipyramidal, three vertices missing
	9	5	*hypo*	cubic, three vertices missing

To work out the total number of skeletal electrons, we must accept that for a transition metal the number of electrons donated to the cluster is $V + X - 12$, where V is the number of valence electrons on the metal, and X is the number of electrons donated from the ligands. In this context V is derived in a similar way to that in which electrons are counted on the metal for the eighteen-electron rule (Section 6.5); the number of electrons, X, is the sum of electrons donated to the metal, and is the same as that in Table 2.

Let us now consider an application of the rules to predict the structure of $Os_5(CO)_{16}$.

☐ Divide $Os_5(CO)_{16}$ into fragments suitable for the Wade–Mingos rules, and then decide on the number of electrons each fragment can donate to the cluster by using the $V + X - 12$ formula.

■ As $Os_5(CO)_{16}$ contains five metal atoms, it can be divided into five structural units. There are sixteen CO ligands and these will be as evenly distributed as possible to form four $Os(CO)_3$ structural units and one $Os(CO)_4$ structural unit. For an $Os(CO)_3$ unit, $V + X - 12 = 8 + 6 - 12 = 2$ skeletal electrons donated to the cluster. For the $Os(CO)_4$ unit, $V + X - 12 = 8 + 8 - 12 = 4$ skeletal electrons donated to the cluster.

☐ What is the total number of skeletal electrons available in $Os_5(CO)_{16}$? On the basis of the number of structural units and Table 3, what would be the shape of the cluster?

■ There is a total of twelve skeletal electrons $(4 + (4 \times 2))$, equal to six skeletal electron pairs. There are five structural units, so n equals five. Table 3 tells us that the $Os_5(CO)_{16}$ structure is based on a polyhedron with five vertices, a trigonal bipyramid (Figure 39), and the structural type classification is *closo*. (Note that these rules do not enable us to predict how many of the CO ligands will be terminal and how many will be bridging.)

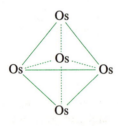

Figure 39 The structural outline of $Os_5(CO)_{16}$ (carbon monoxide ligands are omitted for clarity), as predicted by the Wade–Mingos rules. It has a trigonal bipyramidal geometry with no vacancies. The structural type classification is *closo*.

If the number of electron pairs exceeds the number of structural units by two, the structure opens out, one vertex being unoccupied; the structural type classification is then ***nido*** (nest-like). If the number of electron pairs exceeds the number of structural units by three, two vertices are unoccupied and the structural type classification is called ***arachno*** (web-like). If three vertices are vacant, then the structure is designated as ***hypo***. The classifications are summarised in Table 3, and are illustrated in Figure 40 for the octahedral case.

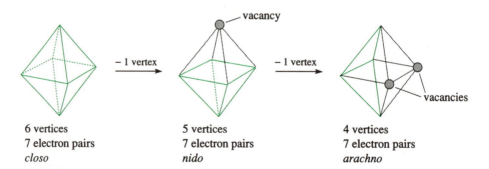

Figure 40 Octahedral geometry with changes in structural units and electron pair counts, as indicated by the Wade–Mingos rules.

☐ Predict the structure of $[FeCo_3(CO)_{12}]^-$ according to the Wade–Mingos rules. Remember that the charge on the complex must be taken into account.

■ $[FeCo_3(CO)_{12}]^-$ contains four metal atoms and so there are four structural units, namely one $Fe(CO)_3$ and three $Co(CO)_3$ units.

For $Fe(CO)_3$ $V + X - 12 = 8 + 6 - 12 = 2$ skeletal electrons.

$Co(CO)_3$ $V + X - 12 = 9 + 6 - 12 = 3$ skeletal electrons.

As there are three Co(CO)$_3$ fragments, a total of nine electrons are donated by Co$_3$(CO)$_9$. There is an overall negative charge on the species, so the total number of skeletal electrons is 12 (2 + 9 + 1), and the total number of electron pairs is six. As n, the number of structural units, is 4, there are $n + 2 = 6$ skeletal electron pairs. Table 3 indicates the structural type classification to be *nido*.

We can now use Table 3 to predict the shape of the [FeCo$_3$(CO)$_{12}$]$^-$ cluster. We know that it has four structural units, and that it has six electron pairs; thus it is based on a trigonal pyramid with one vertex missing (Figure 41).

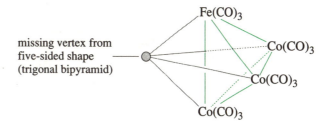

Figure 41 Predicted shape of [FeCo$_3$(CO)$_{12}$]$^-$ on the basis of the Wade–Mingos rules.

☐ Predict the structure of [Os$_3$(CO)$_{12}$] according to the Wade–Mingos rules.

■ [Os$_3$(CO)$_{12}$] has three metal atoms, and so we divide it into three structural units, namely three Os(CO)$_4$ units. For each structural unit the formula gives the number of electrons donated to the cluster as $V + X - 12 = 8 + 8 - 12 = 4$. As there are three structural units, this gives a total of twelve skeletal electrons, and thus six skeletal electron pairs.

The total number of structural units, n, is three and so there are three more electron pairs than structural units. From Table 3 we see that this combination of six skeletal electron pairs and three structural units leads to an *arachno* structural type classification. This means that the structure will be based on a shape having two more vertices than there are metal atoms — that is, five vertices (a trigonal bipyramid) — but two of the vertices will be unoccupied (see Figure 42b, p.58).

So far, we have only applied the Wade–Mingos rules to clusters that only contain transition metals. We can now extend these ideas to mixed clusters, which also contain main-Group fragments, and even to non-metal fragments such as ·C̈H$_2$; ·C̈R, etc.

For non-transition-element structural units the number of electrons donated to the bonding of the cluster is given by $V + X - 2$. For ·C̈H$_2$, $V + X - 2 = 4 + 2 - 2 = 4$ skeletal electrons or two skeletal electron pairs.

☐ How many electrons would the radical ·C̈R donate (where R is an alkyl group)?

■ For ·C̈R, $V + X - 2 = 4 + 1 - 2 = 3$ electrons.

Now consider the mixed cluster Os$_2$(CO)$_8$C̈H$_2$. There are three structural units, comprising one ·C̈H$_2$ unit and two Os(CO)$_4$ units, so $n = 3$.

The total electron count for the Os(CO)$_4$ units is $2 \times (V + X - 12) = 2(8 + 8 - 12) = 8$ electrons. For the main-Group fragment, $V + X - 2 = 4 + 2 - 2 = 4$ electrons available for cluster bonding. Overall, this gives twelve electrons, or six electron pairs, available for cluster bonding. The combination of six electron pairs ($n + 3$) and three structural units is predicted by Table 3 to have an *arachno* structural type classification. The structure will thus be based on the five-vertex trigonal bipyramid, with two vacancies.

Remember that the Wade–Mingos theory does not predict the distribution of terminal and bridging CO groups, or which vertices are absent in a cluster. However, when faced with a large number of diverse possible structures, we are now in a much better position to predict molecular geometry. This is an extraordinarily powerful concept, such that by considering just the number of electrons and the number of structural units we can predict the structures of main-Group (for example carboranes such as $C_2B_4H_6$), transition-metal and mixed clusters.

□ Are $\cdot\dot{C}H_2$ and $Os(CO)_4$ isolobal? Do these fragments donate the same number of electrons under the Wade–Mingos rules? Use your findings to predict the possible formation of two complexes containing these structural units.

■ Yes they are isolobal because $\cdot\dot{C}H_2$ has two half-filled orbitals $(8 - (4 + 2))$ and $Os(CO)_4$ also has two half-filled orbitals $(18 - (8 + 8))$. According to the Wade–Mingos rules, $Os(CO)_4$ donates $V + X - 12 = 8 + 8 - 12 = 4$ electrons. $\cdot\dot{C}H_2$ is a main-Group fragment, so it donates $V + X - 2 = 4 + 2 - 2 = 4$ electrons. Hence, by the Wade–Mingos rules, $\cdot\dot{C}H_2$ and $Os(CO)_4$ both donate the same number of electrons. We already know of the existence of $[Os_3(CO)_{12}]$, so we predict that a $Os(CO)_4$ unit could be replaced with a $\cdot\dot{C}H_2$ unit. $[Os_2(CH_2)(CO)_8]$ and $[Os(CH_2)_2(CO)_4]$ are then possible isostructural compounds.

□ What structures do you predict for $[Os_2(CH_2)(CO)_8]$ and $[Os(CH_2)_2(CO)_4]$?

■ Both clusters have three structural units ($n = 3$), and six skeletal electron pairs, $n + 3$. They are therefore predicted to have *arachno* structural type classifications. Hence their structures will be based on a five-vertex polyhedron (trigonal bipyramid) with two vertices vacant, leading to trigonal structures, as shown in Figure 42(a and c). The same result could have been derived if we knew the structure of any of the clusters in Figure 42: as $Os(CO)_4 \longleftrightarrow \cdot\dot{C}H_2$, we can replace one group with another in the cluster.

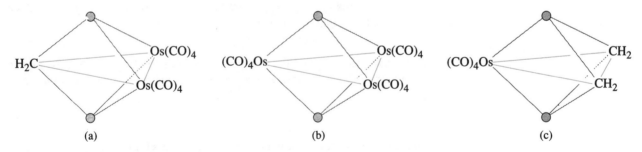

Figure 42 Shapes of (a) $[Os_2(CH_2)(CO)_8]$, (b) $[Os_3(CO)_{12}]$, and (c) $[Os(CH_2)_2(CO)_4]$, predicted on the basis of the Wade–Mingos rules and correlated with the isolobal analogy.

SAQ 11 Use the Wade–Mingos rules to predict the structures of:

(a) $[Co_4(CO)_{12}]$;

(b) $[Rh_6(CO)_{16}]$;

(c) $[Ni_5(CO)_{12}]^{2-}$;

(d) $[Ru(CO)_3Co_2(CO)_6Se]$;

(e) $[cpNiCo_3(CO)_9]$;

(f) B_5H_9;

(g) $C_2B_4H_8$;

(h) $[Ru_2Rh_2H_2(CO)_{12}]$, which you met in Block 5, Section 2 (*hint* H donates one electron).

6.8 The chemistry of alkene complexes

After devoting our attention to the theoretical concepts of the eighteen-electron rule, the isolobal analogy and the Wade–Mingos rules, which enable us to predict cluster and molecular geometry, we shall now consider how the reactivity of transition-metal alkene complexes can in some cases be rationalised.

6.8.1 Synthetic aspects

The generation of alkene complexes by hydride transfer from a coordinated alkyl group was mentioned in Section 3.1 when we considered β-hydrogen elimination (reaction 2). A variation of this process, the abstraction of hydrogen (formally as H$^-$) by chemical methods, provides a convenient route to alkene complexes: for example

$$[Mn(CH_2CH_3)(CO)_5] + Ph_3C^+\,BF_4^- \longrightarrow [Mn(CO)_5(\eta^2\text{-}C_2H_4)]^+\,BF_4^- + Ph_3CH \qquad 66$$

Ph$_3$C$^+$ and R$_3$O$^+$ salts are particularly useful in this context, since they produce by-products (Ph$_3$CH or R$_2$O + RH) that do not impede the isolation of the products.

The most common preparative method for alkene complexes involves displacement of another ligand by an alkene molecule; for example

$$[PdCl_4]^{2-} + C_2H_4 \longrightarrow [PdCl_3(\eta^2\text{-}C_2H_4)]^- + Cl^- \qquad 67$$

$$[FeCl(\eta^5\text{-}C_5H_5)(CO)_2] + C_2H_4 + AlCl_3 \longrightarrow [Fe(\eta^5\text{-}C_5H_5)(CO)_2(\eta^2\text{-}C_2H_4)]^+\,AlCl_4^- \qquad 68$$

Zeise's salt, K[PtCl$_3$(η^2-C$_2$H$_4$)], is prepared by the displacement of chloride ion from [PtCl$_4$]$^{2-}$. The many silver–alkene complexes, formed when alkenes are added to solutions of silver salts, involve the displacement of the solvent or perchlorate, nitrate or tetrafluoroborate ions by alkene molecules. The silver–alkene complexes are labile, and are consequently difficult to isolate.

Displacement of strong-field π-bonded groups, such as CO, may be achieved thermally, but ultraviolet photolysis is often preferred since the reactions tend to proceed more smoothly and are more readily controlled; for example

$$[Mn(\eta^5\text{-}C_5H_5)(CO)_3] + C_2H_4 \xrightarrow{h\nu} [Mn(\eta^5\text{-}C_5H_5)(CO)_2(\eta^2\text{-}C_2H_4)] + CO \qquad 69$$

Another photochemical preparative route is:

$$[Pd(PPh_3)_4] + (NC)_2C{=}C(CN)_2 \xrightarrow{h\nu} [Pd\{\eta^2\text{-}(NC)_2C{=}C(CN)_2\}(PPh_3)_2] + 2PPh_3 \qquad 70$$

Complexes of conjugated dienes may be synthesised by similar routes, starting from the conjugated diene. Also, in some reactions of non-conjugated dienes, rearrangement by hydrogen transfer occurs to produce the conjugated diene derivatives; for example

As we saw for ethene in Section 6.2.1 and Figure 24, the essential effects on an alkene when bonded to a metal include a lowering of the electron density in the π orbital, and an increase of the electron density in the π* orbital. Thus, the overall process may be regarded as an excitation of a π-bonding electron to the π* level. Consequently, the bonded alkene is in an activated and reactive state, and will undergo a wide variety of reactions, many of which are of major industrial importance. Since transition metals are able to activate other small molecules (such as CO, O$_2$, alkynes) in a similar way, complexes of transition metals are very accomplished catalysts. The role of organotransition-metal complexes as homogeneous catalysts is explored in Section 8.

6.8.2 Reactions with electrophiles

In contrast to free alkenes, coordinated alkenes rarely undergo electrophilic addition. This difference in mode of reaction relates to the polarisation of the alkene π-electron cloud towards the metal on coordination. The side of the coordinated alkene opposite the metal becomes more susceptible to attack by nucleophiles. In reactions 72–4 we show examples of electrophilic reactions involving direct protonation and hydrogen abstraction, which result in a change in the alkene ligand and its mode of attachment to the metal:

$$\text{(butadiene)}Fe(CO)_3 + HCl \longrightarrow \text{(methylallyl)}Fe(CO)_3Cl \qquad 72$$

$$\text{(pentadienyl)}Fe(CO)_3 + Ph_3C^+ \longrightarrow [\text{(pentadienyl)}Fe(CO)_3]^+ + Ph_3CH \qquad 73$$

$$\text{(cod)(cp)Co} + Ph_3C^+ \longrightarrow [\text{(cyclooctadienyl)(cp)Co}]^+ + Ph_3CH \qquad 74$$

In reaction 72, electrophilic protonation of the η^4-butadiene ligand yields a product containing a η^3-C_4H_7 ligand. In reaction 73, hydrogen abstraction from the methyl group of the η^4-penta-1,3-diene ligand converts it to a η^5-C_5H_7 ligand. In reaction 74, hydrogen abstraction from cyclo-octadiene changes the *hapticity* of the ligand from η^4 to η^5.

6.8.3 Substitution reactions

Alkene complexes provide a ready access (via alkene displacement) to other metal complexes, and have found wide use as starting materials. Notable examples of such complexes are $[Mo(CO)_4(\eta^4\text{-norbornadiene})]$ and $[Fe(\eta^4\text{-}C_4H_6)(CO)_3]$, as illustrated in reactions 75 and 76:

norbornadiene

$$[Mo(CO)_4(\eta^4\text{-norbornadiene})] + 2PPh_3 \longrightarrow [Mo(CO)_4(PPh_3)_2] + \text{norbornadiene} \qquad 75$$

$$[Fe(\eta^4\text{-}C_4H_6)(CO)_3] + 2PPh_3 \longrightarrow [Fe(CO)_3(PPh_3)_2] + C_4H_6 \qquad 76$$

6.8.4 Nucleophilic addition to polyenes — Green's rules

The d-block metals have a strong electron demand, and so coordinated polyenes are more susceptible to nucleophilic attack than the free hydrocarbon. The mode of attack may be either *exo* or *endo* in nature; that is, it can occur either from the side opposite the metal (*exo*) or from the same side (*endo*).

$$\text{exo product} \xleftarrow{exo} [\text{cation} + Nu^-] \xrightarrow{endo} \text{endo product} \qquad 77$$

In the *endo* case some form of metal-assisted coordination normally occurs. In both cases, the addition causes the ligand to reduce its *hapticity* by one.

The site of nucleophilic attack can be worked out by the step-by-step application of **Green's rules**[†] for cationic metal species. (Note that rule 1 is the most important, followed by rule 2, followed by rule 3.)

1 Even ligands are attacked in preference to odd.

2 Open ligands are attacked in preference to closed.

3 For an open even polyene, the terminal carbon is preferentially attacked. (For odd open polyenes the terminal carbon is only attacked if the remainder of the metal–ligand system is strongly electron withdrawing.)

Even ligands are those with an even number of conjugated atoms (η^2, η^4, η^6, etc.); *odd ligands* are those with an odd number of conjugated atoms (η^1, η^3, η^5, etc.).

Closed means cyclically conjugated, whereas *open* indicates non-cyclically conjugated. It must be noted that this set of rules predicts the favoured position of attack on *kinetic* grounds. Typical nucleophiles can be $^-$CN, $^-$H and $^-$OMe. Green's rules are best explained by reference to the examples shown in Figure 43.

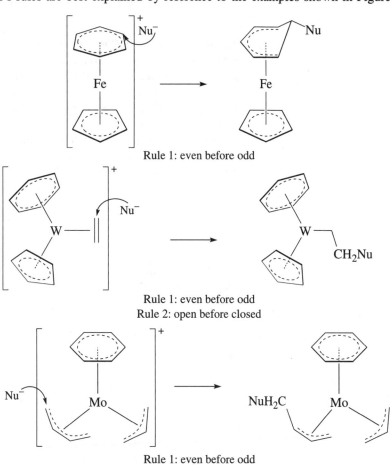

Figure 43 Examples of the application of Green's rules. (Nu$^-$ is a nucleophile such as $^-$CN, $^-$OMe, $^-$H.) Note that rule 1 (even before odd) has priority over rule 2 (open before closed), which has priority over rule 3 (terminal carbon of even open polyene preferentially attacked).

The rules can be derived from a consideration of the electron distribution. An even polyene has two electrons in its **highest-occupied molecular orbital** (HOMO) — that is, the highest-energy molecular orbital containing an electron(s) — whereas an odd polyene has one electron in its HOMO.

[†] Named after Professor M. Green.

☐ On the basis of the simplified molecular orbital diagrams for η^4-butadiene, η^3-allyl and η^2-ethene below, identify the HOMO orbital and note the number of electrons in it.

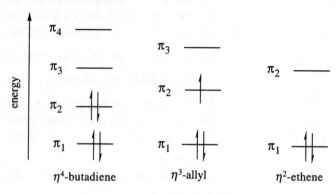

■ In η^4-butadiene the HOMO is π_2 and it has two electrons.

In η^3-allyl the HOMO is π_2 and it has one electron.

In η^2-ethene the HOMO is π_1 and it has two electrons.

Assuming that the HOMO is the major donor orbital, an even polyene has a greater partial positive charge than an odd polyene, since more electrons have been donated in coordination to the metal; this is the basis for rule 1. Cyclic systems are able to delocalise charge evenly around themselves, whereas open polyenes are not able to delocalise the charge so effectively. The uneven electron distribution in open systems results in sites that are partially positively charged and hence are susceptible to nucleophilic addition; this is the basis for rule 2. For even open polyenes the charge (δ^+) tends to localise itself on the terminal carbons; this is the basis for rule 3.

☐ Predict the sites of nucleophilic addition in the polyenes shown in Figure 44.

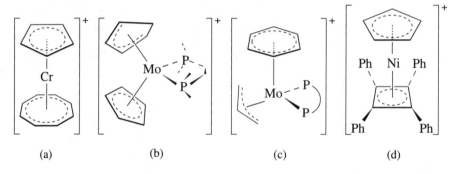

Figure 44 Further examples of the application of Green's rules.

■ (a) Rule 1 (even before odd) predicts addition on the η^8 ring. Note that for cyclic conjugated systems without substituents all ring positions are equivalent, and therefore are equally susceptible to addition.

(b) Rule 1: predicts addition on the η^4 ring (note that the η^4-coordinated five-membered ring is regarded as an open conjugated system because the four atoms forming the delocalised system in the upper C_5 ring do not form a closed ring).

(c) Rule 1 predicts addition on the η^6 ring.

(d) Rule 1 predicts addition on the η^4 ring.

SAQ 12 If two attacks by a nucleopile Nu⁻ occurred consecutively in the following complexes, what would be the sites of addition and what would be the products?

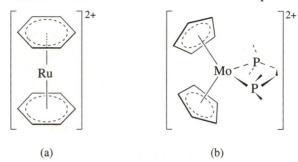

(a)   (b)

SLC 8   Green's rules provide us with great predictive power to determine the site of nucleophilic addition on cationic polyene systems. In this way organometallic chemistry shows its heritage between organic chemistry, where *regioselectivity* of attack is well defined, and inorganic chemistry, where the complex nature of the subject make application of rules more difficult.

6.9 Summary of Section 6

1 Alkenes bond to transition metals through a combination of σ and π orbital interactions; they can be regarded as η^2 ligands. For alkenes, and related ligands such as the η^3-allyl group, the metal is positioned above the plane of the carbon–carbon double bond.

2 Coordinated alkenes, in contrast to free alkenes, are readily susceptible to nucleophilic addition, but are less susceptible to attack by electrophiles.

3 The eighteen-electron rule can be used to predict the formulae of complexes; knowing the electron contribution to the bonding of the other ligands, the electron contribution from an organo group or molecule, and its likely *hapticity*, can be determined.

4 Two molecular fragments are isolobal if the number, symmetry properties, approximate energy and shape of their valence orbitals, as well as the number of electrons occupying them are similar. The isolobal analogy enables us to predict which molecular fragments are potentially able to combine together to form complexes and clusters.

5 The Wade–Mingos rules enable us to predict the shape of a cluster of metal atoms by considering the difference between the number of skeletal electron pairs and the number of structural units. This leads to *closo*, *arachno*, *nido* and *hypo* clusters, dependent on the number of structural unit vacancies.

6 The site of nucleophilic addition on cationic polyene systems can be predicted using Green's rules. Even systems are attacked in preference to odd, open systems are attacked in preference to closed, and the terminal carbon of open even systems is preferentially attacked.

7 CYCLIC POLYHAPTO SYSTEMS

In this Section we shall extend our study of organometallic chemistry to include more complex compounds containing polyhapto ligands. We shall also extend the ideas of bonding we developed in the last Section to illustrate the diversity of *polyhapto* organometallic chemistry.

7.1 Structure and bonding in η^5-cyclopentadienyl complexes

The development of transition-metal organometallic chemistry, although promoted by the report of $K[PtCl_3(C_2H_4)]$ by W. C. Zeise in 1831, of $[Ni(CO)_4]$ by L. Mond in 1890, and of polyphenylchromium complexes by F. Hein in 1919, obtained an innovative advance by the report of *bis*(cyclopentadienyl)iron (trivial name 'ferrocene') in 1951. Although the nature of the bonding of the cyclopentadienyl group to the iron was not immediately appreciated by the authors, T. J. Kealy and P. L. Pauson, their discovery was undoubtedly one of the most important events to stimulate subsequent advances in organometallic chemistry. The structure proposed by Kealy and Pauson of a σ complex (structure **3**) was quickly disputed by G. Wilkinson and R. Woodward, who suggested that ferrocene was a *polyhapto* π-bonded **sandwich compound**, otherwise known as a *metallocene*. The single line shown in the ^{1}H n.m.r. spectrum for the ring protons was good evidence for an aromatic structure of the $C_5H_5^-$ ligand, with all the carbon atoms equidistant from iron. In 1956 the sandwich structure was confirmed by X-ray crystallography. In the solid state, its planar *pentahapto* cyclopentadienyl groups, originally thought to have a staggered conformation, are now known to be twisted only about 9° from an eclipsed conformation (Figure 45). Related metallocenes of Ru, Os, V, Cr, Co and Ni have similar structures. The barrier to rotation of the rings is small, and in solution the separate conformations cannot be detected, even at very low temperatures.

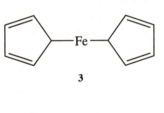

3

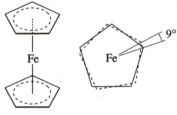

Figure 45 Two views of the molecular structure of ferrocene in the solid state.

Ferrocene has a number of unusual properties; it is diamagnetic and, unlike most other sandwich compounds, it is air stable. It undergoes electrophilic substitution on its C_5H_5 rings much faster than does benzene and it is easily oxidised both electrochemically and chemically by, for example, iodine. Thus, a bonding description is required for ferrocene which can account for all these properties.

Consider the π orbitals of the cyclopentadienyl groups which are suitable for bonding to the metal. In addition to the σ-bonding framework, each ring carbon atom has one p_z orbital and one electron available for π-bonding. Thus, from five atomic orbitals, five molecular orbitals may be constructed. Suitable combinations of cyclopentadienyl and iron orbitals are shown in Figure 46; their energies increase from bottom to top.

Interaction with the iron orbitals having the appropriate symmetry for overlap leads to iron–cyclopentadiene σ-bonding through π_1, leads to π-bonding through π_2 and π_3, and δ-bonding through π_4 and π_5. The simplified molecular orbital energy-level diagram for ferrocene that results from such considerations, which takes into account the interactions of two cyclopentadienyl rings, is shown in Figure 47. Since each ring contributes five electrons to the bonding and the metal contributes eight electrons, levels up to, and including, the n' orbital are filled. The n' orbital is predominantly iron $3d_{z^2}$ in character and constitutes the HOMO. The next highest level, the 2δ (d_{xy}, $d_{x^2-y^2}$) molecular orbitals, are only slightly bonding in character and retain predominantly $3d_{xy}$ and $3d_{x^2-y^2}$ metal character. Thus, the three molecular orbitals of highest energy, accommodating six electrons, are essentially non-bonding iron orbitals. Important bonding molecular orbitals are those arising from the interaction of $3d_{xz}$ and $3d_{yz}$ (and, to a much smaller extent, $4p_x$ and $4p_y$) iron orbitals with the π_2 and π_3 sets of cyclopentadienyl orbitals (Figure 46).

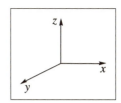

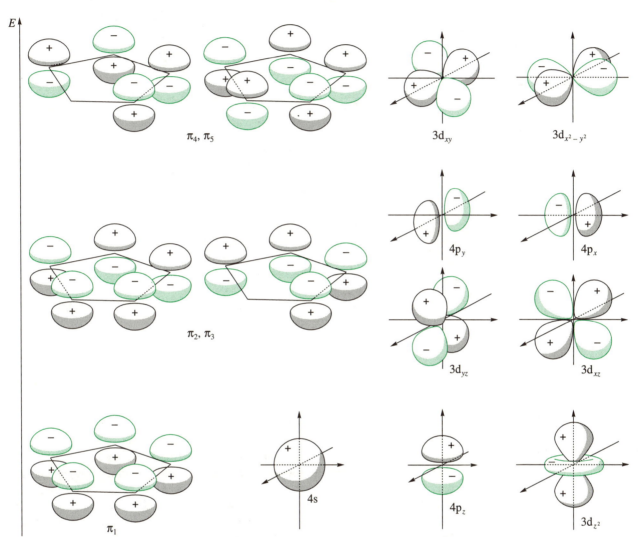

Figure 46 Sets of orbitals for ferrocene: iron atomic orbitals of appropriate symmetry are matched with the appropriate molecular orbitals on the cyclopentadienyl ring.

☐ What is the electron count for [Cr(C$_5$H$_5$)$_2$], [Fe(C$_5$H$_5$)$_2$] and [Co(C$_5$H$_5$)$_2$]. Use the molecular orbital energy-level diagram for ferrocene (Figure 47) to predict the number of unpaired electrons in [Cr(C$_5$H$_5$)$_2$] and [Co(C$_5$H$_5$)$_2$], and hence determine the magnetic moment of these complexes.

■ [Cr(C$_5$H$_5$)$_2$] = 6 + 5 + 5 = 16 electrons

[Fe(C$_5$H$_5$)$_2$] = 8 + 5 + 5 = 18 electrons

[Co(C$_5$H$_5$)$_2$] = 9 + 5 + 5 = 19 electrons

[Cr(C$_5$H$_5$)$_2$] could either be diamagnetic (no unpaired electrons) or paramagnetic, with $\mu \approx 2.83\mu_B$ (two unpaired electrons), depending on the energy difference between the n' and 2δ levels and the spin pairing energy†. [Fe(C$_5$H$_5$)$_2$] has no unpaired electrons, so is diamagnetic (magnetic moment of zero). [Co(C$_5$H$_5$)$_2$] has one unpaired electron, so would be paramagnetic, with $\mu \approx 1.73\mu_B$.

† The empirical value for [Cr(C$_5$H$_5$)$_2$] is close to $2.83\mu_B$, indicating that the energy difference between the n' and 2δ levels is small compared to the spin pairing energy.

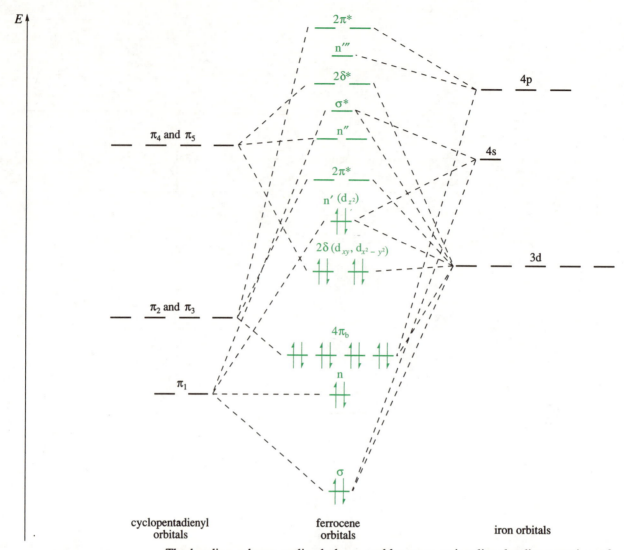

Figure 47 Molecular orbital energy-level diagram for ferrocene.

The bonding scheme outlined above enables us to rationalise the diamagnetism of ferrocene. Moreover, the six non-bonding electrons (n' and 2δ levels) reside mainly in iron 3d orbitals, and the lower levels retain predominantly cyclopentadienyl orbital character (their energies are closer to those of the cyclopentadienyl orbitals (π_1) than to the iron atomic orbital (4s, $4p_z$ and $3d_{z^2}$)). This leads to the conclusion that the metal is present as iron(II) with a d^6 configuration, and that it has a positive charge relative to the rings. Thus, the complex may be considered as two aromatic cyclopentadienide ($C_5H_5^-$) rings bonding to an iron(II) centre. This type of electron distribution is confirmed by the ease with which ferrocene undergoes electrophilic aromatic substitution at the ring. The basic bonding scheme also applies to other metallocenes, although the relative energies of the levels will change with the different energies of the metal atomic orbitals and the number of electrons in the complexes. Changing the number of electrons in the upper two occupied levels is not expected to have a major effect on the bonding of the cyclopentadienyl groups to a metal atom, and indeed the metallocenes of titanium, vanadium and chromium (neutral or cationic) are well known.

As you may have gathered from the question above, most metallocenes are paramagnetic, their magnetic properties being interpreted readily in terms of the distribution of all electrons in excess of twelve within the 2δ, n' and 2π* levels (essentially non-bonding metal orbitals). The cobalticenium ion, $[Co(C_5H_5)_2]^+$, which is isoelectronic with ferrocene (both eighteen-electron species) is particularly robust, but $[Co(\eta^5\text{-}C_5H_5)_2]$, $[Ni(\eta^5\text{-}C_5H_5)_2]$ and $[Ni(\eta^5\text{-}C_5H_5)_2]^+$ are unstable relative to ferrocene, since some electrons have to occupy antibonding metallocene orbitals (Figure 47). Indeed nickelocene, because of its lability, is widely used as a starting material for other nickel complexes, and cobaltocene is readily oxidised to yield the eighteen-electron cobalticenium cation. Sandwich compounds, including lanthanide and actinide metallocenes, do not, in general, follow the eighteen-electron rule. However, the eighteen-electron species [Fe(C$_5$H$_5$)$_2$], [Co(C$_5$H$_5$)$_2$]$^+$ and [Cr(C$_6$H$_6$)$_2$] are very stable.

7.2 Other cyclic *polyhapto* ligands

So far we have rationalised the existence of *monohapto* complexes (Section 5) and discussed the *polyhapto* bonding of cyclopentadienyl groups in ferrocene and other metallocenes (Section 7.1). Besides these two arbitrary *hapto* divisions, there are also metallocyclic complexes in which the metal forms part of a saturated ring such as in structures **4** and **5**. These complexes are normally formed by metathesis of dilithium reagents with metal chloride salts. Recently, metallocyclic complexes have been studied to evaluate the mechanisms of organometallic reactions, especially in C—H activation and catalysis.

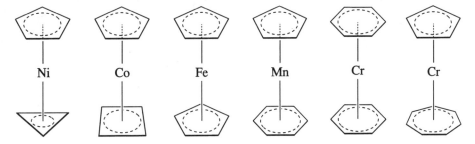

(M = Ni, Pt, Pd; L = PPh$_3$, PMe$_3$; x = 2, 3, 4)

A wide range of sandwich complexes which contain cyclic *polyhapto* ligands and obey the eighteen-electron rule is known, and is illustrated in Figure 48. The ring sizes can vary from η^3-C$_3$H$_3^-$ to η^8-C$_8$H$_8^{2-}$, and even larger.

Figure 48 Some examples of sandwich compounds that obey the eighteen-electron rule.

The complexes in Figure 48 have all been synthesised, and have the general formula [M(C$_n$H$_n$)(C$_x$H$_x$)].

It is also possible for cyclic *polyhapto* ligands to occur in half-sandwich (Figure 49a), multidecker (Figure 49b) and tilted sandwich (Figure 49c) structures. A particularly interesting class is the mixed carborane–metallocene complexes (Figure 49d), in which metals are attached to the polyene and also to the carborane fragment. They are formed by reduction of the carborane with an alkali metal followed by addition of a suitable metal polyenyl anion. Linked metallocenes are also known (Figure 49e).

Apart from (η^5-C$_5$H$_5$) complexes like ferrocene, [(Fe(C$_5$H$_5$)$_2$)], the most extensively studied sandwich compounds are the η^6-arene complexes. The most stable of these is the eighteen-electron dibenzenechromium [Cr(η^6-C$_6$H$_6$)$_2$]. Such η^6-arene complexes have similar properties to the η^5-metallocenes (Section 7.4) and are known for most of the transition metals. Space limitations will restrict our discussion of this important class of *bis*(arene) metal complexes.

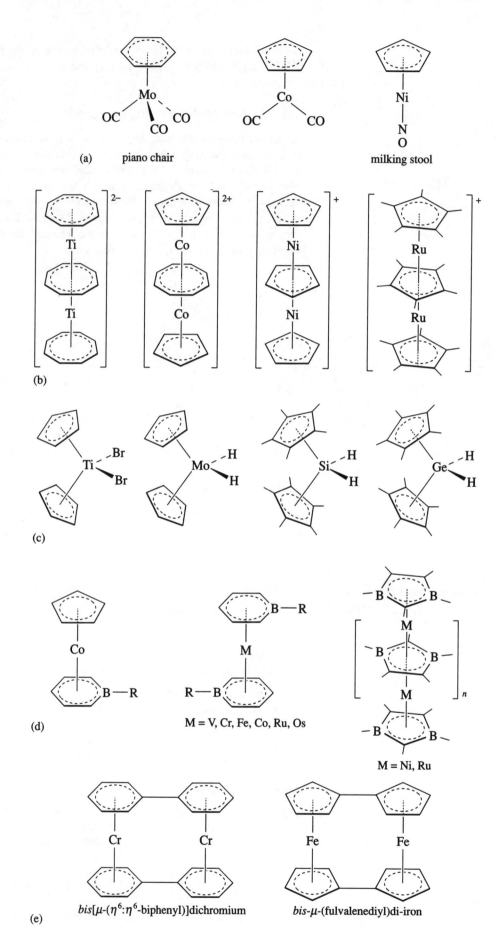

Figure 49 Examples of various types of complex containing cyclic *polyhapto* ligands: (a) half-sandwich structures; (b) multidecker polyenes; (c) tilted sandwich structures; (d) mixed carborane metallocene complexes; (e) linked polyenes.

The bonding in complexes containing cyclic *polyhapto* ligands can be deduced by matching metal orbitals of the correct symmetry with those of the polyene, similar to the treatment of butadiene and ferrocene (Sections 6.2.2 and 7.1).

A selection of the polyenes that are important in bonding to transition metals is shown in Figure 50.

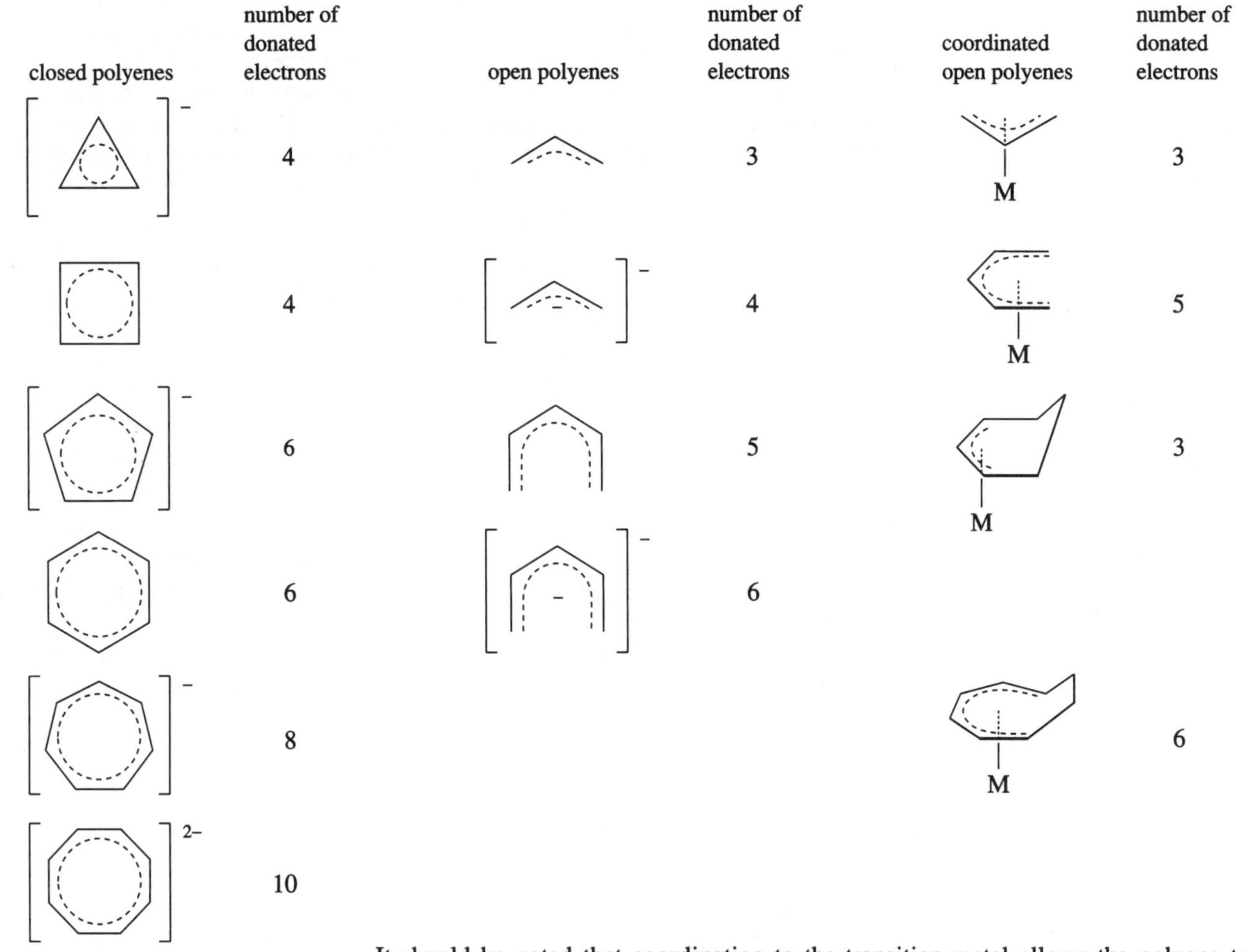

closed polyenes	number of donated electrons	open polyenes	number of donated electrons	coordinated open polyenes	number of donated electrons
△⁻	4	∧	3	∨–M	3
□	4	[∧]⁻	4	◠–M	5
[⬠]⁻	6	⌒	5	⌒–M	3
⬡	6	[⌒]⁻	6		
[⬣]⁻	8			⌒–M	6
[⬢]²⁻	10				

Figure 50 Polyenes, open and closed, that are important in transition-metal organometallic chemistry, and the number of electrons they donate to the metal.

It should be noted that coordination to the transition metal allows the polyene to act as a Lewis base by donating electron density from the π orbital to the metal. In so doing, fragments such as cyclobutadiene which have not been isolated in the free state—except at the temperature of liquid nitrogen (77 K)—can be formed. Furthermore, the same cyclic polyene ligands can act as *monohapto* or *polyhapto* ligands depending on the particular electronic and steric demands of the complex.

SAQ 13 Draw diagrams to suggest alternative ways in which the cyclohexadienyl group (C_6H_7) could bond to a transition metal.

7.3 Range of derivatives of cyclic *polyhapto* ligands

Although cyclic *polyhapto* ligands principally form complexes with the d-block transition metals, *polyhapto* derivatives are also known for the main-Group metals. Sodium and thallium cyclopentadienyls, for example, both have *pentahapto* bonding interactions to the metals, but the bonding is essentially ionic in character. Sandwich-type complexes are also formed by both lanthanide and actinide elements, and many complexes of the two series are isostructural (Figure 51).

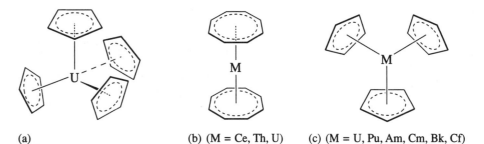

Figure 51 A range of metallocenes for actinide and lanthanide elements with cyclic *polyhapto* ligands:
(a) $[U(\eta^5\text{-}C_5H_5)_4]$; (b) $[M(\eta^8\text{-}C_8H_8)_2]$; (c) $[M(\eta^5\text{-}C_5H_5)_3]$.

(a) (b) (M = Ce, Th, U) (c) (M = U, Pu, Am, Cm, Bk, Cf)

The actinides form three series of derivatives containing the η^5-C_5H_5 ligand: (i) [M(η^5-C_5H_5)$_3$], (ii) [M(η^5-C_5H_5)$_4$] and (iii) [MX(η^5-C_5H_5)$_3$], where X is a halide or other anionic group. Tetrahedral arrays of the η^5-C_5H_5 rings (and halide where appropriate) about the metal are found for all the complexes (see, for example, Figure 51a). For the [M(η^5-C_5H_5)$_3$] complexes (Figure 51c), a ring associated with a neighbouring unit at a distance somewhat larger than that to the other three rings, completes the tetrahedral arrangement. Not surprisingly, these [M(η^5-C_5H_5)$_3$] complexes show Lewis acid properties, and form 1:1 adducts with isonitriles, amines and ethers, the interaction to the essentially bridging fourth ring system being replaced by coordination to the ligand.

The actinides in particular are sufficiently large for their atomic orbitals to overlap effectively with the molecular orbitals of a cyclo-octatetraene ring, and several are known — for example, the sandwich compounds [U(η^8-C_8H_8)$_2$] and [Th(η^8-C_8H_8)$_2$] (see Figure 51b) — in which the planar aromatic ring acts as an *octahapto* ligand. The most interesting feature of the bonding is the involvement of metal f orbitals. Indeed, participation of 5f orbitals may be more widespread than for these cyclo-octatetraene complexes alone: they are also thought to contribute to the bonding in [U(η^5-C_5H_5)$_4$].

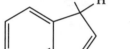

indenide anion

indene

In contrast to the actinides, the lanthanide atomic orbitals (especially the 4f shell) are too contracted and localised to overlap satisfactorily with cyclic *polyhapto* ligands. A series of complexes of formulation [Ln(C_5H_5)$_3$], [LnCl(C_5H_5)$_2$] and [LnCl$_2$(C_5H_5)] has been prepared by reaction of anhydrous LnCl$_3$ (Ln = lanthanide element) and NaC$_5$H$_5$ in tetrahydrofuran. The bonding in these complexes is almost completely ionic. They can consist of tetramers and polymeric species. Lanthanide complexes with the $C_8H_8^{2-}$ and indenide (anion of the hydrocarbon indene), $C_9H_7^-$, ligands are also known.

7.4 The chemistry of metallocenes

Except for their structures, probably the most striking feature of the metallocenes is the ability of a few of them to undergo electrophilic aromatic substitution reactions. Early investigation indicated that ferrocene was a new type of aromatic system, which readily underwent a range of typical electrophilic aromatic substitution reactions, including alkylation, arylation, metallation and sulphonation.

Although many metallocenes (see Figure 48) like ferrocene are known to have aromatic character, most, like nickelocene and cobaltocene, are not capable of undergoing aromatic substitution because they are either oxidised or undergo addition reactions more readily. Consequently, the chemistry of metallocenes which is discussed below is best illustrated by the reactions of ferrocene.

7.4.1 Synthetic aspects

Most preparative methods use cyclopentadiene to generate the cyclopentadienyl group, either by direct reaction or via the intermediate formation of the sodium salt or Grignard derivative, as in reactions 78–81:

$C_5H_6 + FeCl_2 \xrightarrow{Et_2NH} [Fe(\eta^5\text{-}C_5H_5)_2]$ **78**

$C_5H_5MgBr + [Ru(acac)_3] \longrightarrow [Ru(\eta^5\text{-}C_5H_5)_2]$ **79**

$C_5H_5Na + MnBr_2 \xrightarrow{liq.\ NH_3} [Mn(\eta^5\text{-}C_5H_5)_2]$ **80**

$VCl_4 + C_5H_5MgBr \longrightarrow [V(\eta^5\text{-}C_5H_5)_2]$ **81**

Cyclopentadienyl complexes containing other ligands such as carbonyl, hydrido and halides may also be synthesised by similar routes; for example

$TiCl_4 + NaC_5H_5 \longrightarrow [TiCl_2(\eta^5\text{-}C_5H_5)_2]$ **82**

$[Fe(CO)_5] + NaC_5H_5 \longrightarrow [\{Fe(CO)_2(\eta^5\text{-}C_5H_5)\}_2]$ **83**

$ReCl_5 + NaC_5H_5 \longrightarrow [ReH(\eta^5\text{-}C_5H_5)_2]$ **84**

The solvent provides the hydrido ligand in the last example.

7.4.2 Physical aspects

Magnetic measurements show that metallocenes, with certain diamagnetic exceptions including $[Fe(\eta^5\text{-}C_5H_5)_2]$, $[Co(\eta^5\text{-}C_5H_5)_2]^+$ and $[Ti(\eta^5\text{-}C_5H_5)_2]$, are paramagnetic and have electronic configurations with either partially filled n' and 2δ levels or higher levels (such as 2π*; see Figure 47) in the case of cobalt and nickel derivatives.

The cyclopentadienyl ring produces a single resonance peak in the ^{1}H n.m.r. spectrum, except when it is split by coupling with a metal nucleus such as ^{195}Pt ($I = \frac{1}{2}$) or ^{103}Rh ($I = \frac{1}{2}$), or with protons directly attached to the metal, for example $[Re(\eta^5\text{-}C_5H_5)_2H]$. For unsymmetrical cyclopentadienyl complexes, the exchange process involving the interchange of positions of the ring hydrogens is fast (10^{-2}–10^{-9} s) when compared to the nuclear magnetic resonance time-scale, and a time-average signal is detected. This arises from **ring-whizzing** about the metal–ring axis, as illustrated in Figure 52. Even cooling to temperatures below −77 °C fails to slow the process down sufficiently for individual protons to be detected.

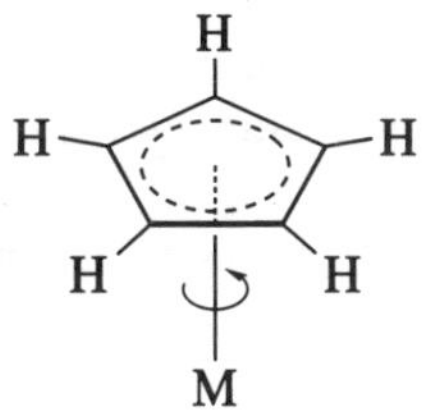

Figure 52 Ring-whizzing of a cyclopentadienyl group about the metal–ligand bond.

Infrared spectroscopy has been of particular use in the detection of unsubstituted cyclopentadienyl rings, a problem that is of importance following attempted ring-substitution reactions of metallocenes. One or both rings may become substituted in these reactions. A rule, known as the **1 100–1 000 cm^{-1} rule** is helpful. It is based on the observation that two absorption peaks occur around this region for unsubstituted rings, for example at 1 104 and 1 001 cm^{-1} for ferrocene, but are absent when the ring is substituted. Such a rule may be used to differentiate between the two isomers in Figure 53, though an alternative method — considered to be more sensitive — involves inspection of the mass spectrum, where the $[M(C_5H_5)]^+$ ion is a fragment invariably present for complexes with unsubstituted rings.

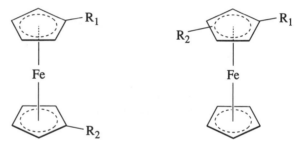

Figure 53 Isomeric disubstitution products of ferrocene.

7.4.3 General properties and reactions

Neutral metallocene complexes are all coloured, volatile solids, the majority of which are very air sensitive, particularly in solution. Ferrocene, however, is stable in air and is thermodynamically the most stable metallocene towards dissociation: the stability order is Fe > Ni > Co > V >> Cr > Ti. Consequently, ferrocene has been used extensively in studies of the chemistry of metallocenes, and figures frequently in the following examples, which illustrate the various types of reaction.

(a) Electron-transfer reactions

The facile oxidation of cobaltocene to the cobalticenium ion has been mentioned previously, but such a process can be achieved for each metallocene (often reversibly); for example

$[Fe(\eta^5\text{-}C_5H_5)_2]^+ + e = [Fe(\eta^5\text{-}C_5H_5)_2]$ $E^\ominus = -0.30\text{ V}$ **85**

$[Co(\eta^5\text{-}C_5H_5)_2]^+ + e = [Co(\eta^5\text{-}C_5H_5)_2]$ $E^\ominus = -1.16\text{ V}$ **86**

$[Ni(\eta^5\text{-}C_5H_5)_2]^+ + e = [Ni(\eta^5\text{-}C_5H_5)_2]$ $E^\ominus = -0.21\text{ V}$ **87**

Ferrocene is easily and rapidly oxidised by, for example, iodine, to form the ferricenium ion. Indeed the ferrocene–ferricenium redox couple is often used as an electrochemical calibration standard.

(b) Protonation of the metal

Up to three hydrogens can be attached to a metallocene using the n' and 2δ orbitals for bonding, as in the complexes [Re(η^5-C_5H_5)$_2$H], [Mo(η^5-C_5H_5)$_2$(H)$_2$] and [Ta(η^5-C_5H_5)$_2$(H)$_3$]. Isoelectronic complexes may be obtained via protonation reactions as follows:

$$[Fe(\eta^5\text{-}C_5H_5)_2] + H^+ \longrightarrow [Fe(\eta^5\text{-}C_5H_5)_2H]^+ \qquad 88$$

$$[Re(\eta^5\text{-}C_5H_5)_2H] + H^+ \longrightarrow [Re(\eta^5\text{-}C_5H_5)_2(H)_2]^+ \qquad 89$$

$$[Mo(\eta^5\text{-}C_5H_5)_2(H)_2] + H^+ \longrightarrow [Mo(\eta^5\text{-}C_5H_5)_2(H)_3]^+ \qquad 90$$

Ferrocene is a very weak base, but it can be readily protonated by a strong acid (reaction 88). In the protonated species, the η^5-cyclopentadienyl rings are tilted with respect to each other, away from the coordinated hydrogen. This reaction is somewhat unusual because attack by other electrophiles occurs at the rings rather than at the metal.

(c) Ring transfer

The reactions of particular interest involve the transfer of a η^5-C_5H_5 group between metals; examples are shown in reactions 91–93

$$[Ni(\eta^5\text{-}C_5H_5)_2] \xrightarrow{[Ni(CO)_4]} [Ni(\eta^5\text{-}C_5H_5)_2(CO)]$$

$$\xrightarrow{[Fe(CO)_5]} [(\eta^5\text{-}C_5H_5)Ni(CO)Fe(CO)_2(\eta^5\text{-}C_5H_5)] \qquad 91$$

$$Li(\eta^1\text{-}C_5H_5) \xrightarrow{[W(CO)_6]} Li[W(\eta^5\text{-}C_5H_5)(CO)_3] \qquad 92$$

$$[M(\eta^5\text{-}C_5H_5)_2] + FeCl_2 \longrightarrow [Fe(\eta^5\text{-}C_5H_5)_2] + MCl_2 \qquad 93$$

$$(M = Mn, Cr, V)$$

When M = Co or Ni in reaction 93, no transfer of η^5-C_5H_5 occurs. Thus, on refluxing [Co(η^5-C_5H_5)$_2$] with iron(II) chloride in tetrahydrofuran solution, metallic iron is deposited, and [Co(η^5-C_5H_5)$_2$]Cl is left in solution. The differing results are a function of the relative redox potentials of iron and the metal M.

(d) Metallation

These reactions illustrate an aspect of the widely studied organic chemistry of metallocenes. Ring hydrogens may be replaced by metals, to provide a convenient means of introducing other functional groups into the ring as shown in Figure 54.

(e) Electrophilic aromatic substitution reactions

The terms 'metallocene', 'ferrocene', 'chromacene' ([Cr(η^6-C_6H_6)$_2$]), etc., were coined to convey the aromatic character of the coordinated polyenyl rings and their similarity to benzene. The ability of ferrocene to undergo Friedel–Crafts acylation reactions was readily appreciated soon after its discovery, and since then a very extensive organic chemistry of ferrocene and other metallocenes has developed.

Friedal–Crafts acylation of ferrocene using acid anhydrides or acyl halides in the presence of BF_3 or $AlCl_3$, is very rapid indeed, the reaction proceeding about 10^6 times faster than for benzene. If the substituent is electron-withdrawing, the disubstituted ferrocenes that result tend to be the isomer with an acyl substituent on each ring rather than one with both substituents on the same ring. Disubstitution at the 3-position of one ring is usually promoted by electron-donating substituents.

Not all electrophilic aromatic substitution reactions that would be expected for the free ligand are achieved for ferrocene because of the ease with which ferrocene oxidises. For example, direct halogenation and nitration are not possible, leading instead to the ferricenium ion. However, halogen and nitro derivatives may be obtained by indirect routes from other substituted ferrocenes. On the other hand, direct sulphonation is possible, using chlorosulphonic acid, for example, as the sulphonating agent.

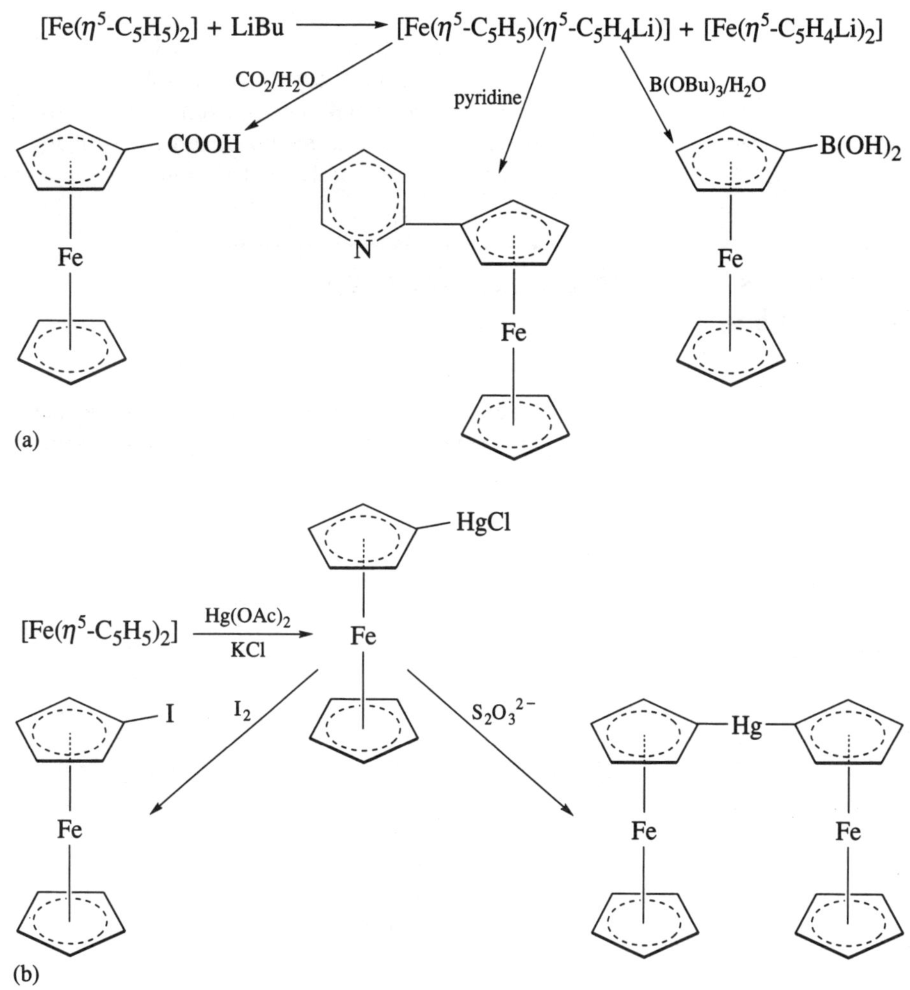

Figure 54 Replacement of ring hydrogens as a route to introducing other functional groups into the cyclopentadienyl rings of ferrocene: (a) reactions involving lithium substitution; (b) reactions involving mercuration.

You will have the opportunity of studying ferrocene chemistry at the CHEM 777 Summer School.

SAQ 14 (a) What is the electron count for the metal in nickelocene, $[Ni(C_5H_5)_2]$?

(b) Is nickelocene paramagnetic or diamagnetic? Use the molecular orbital energy-level diagram (Figure 47) to predict the number of unpaired electrons and hence the approximate magnetic moment.

(c) Suggest a way of synthesising $[Ni(C_5H_5)_2]$.

(d) Suggest two techniques for demonstrating that nickelocene contains unsubstituted cyclopentadienyl rings.

(e) List up to five chemical reactions that you would expect nickelocene to undergo.

7.5 Summary of Section 7

1 The 'sandwich' structure of ferrocene arises from σ, π, and δ interaction of iron atomic orbitals with the five π molecular orbitals of the cyclopentadienyl group. The electron distribution in the resulting molecules shows that six electrons, in the highest filled levels, are essentially located on iron as non-bonding electrons. This provides an explanation for the existence of a whole series of metallocenes of the same general formula $[M(\eta^5-C_5H_5)_2]$, in which parallel cyclopentadienyl rings sandwich a metal atom. The number of electrons in these highest levels has little effect on the stability of the metallocenes. Rings of three carbon atoms up to rings of eight carbon atoms, as well as heterocyclic rings, act as *polyhapto* ligands to metals, and exhibit typical aromatic properties. Similar principles to those used for η^5-C_5H_5 derivatives can be used to describe the bonding of other *polyhapto* ligands to metals.

2 Metallocenes are fluxional in solution as a result of ring-whizzing about the metal–ring axis. They undergo a series of electron transfer, protonation, ring cleavage, metallation and substitution reactions, many typical of aromatic molecules.

Indeed, the aromaticity of the η^5-cyclopentadienyl group is a particularly significant feature of metallocenes. The $1100-1000\,\text{cm}^{-1}$ rule is helpful in distinguishing between substituted and unsubstituted C_5H_5. Apart from ferrocene itself, most neutral metallocenes from the first-row transition elements are paramagnetic. The synthetic methods, and the structural and bonding features of η^6-arene complexes, are closely related to those of η^5-cyclopentadienyl complexes.

8 CATALYSIS BY TRANSITION-METAL COMPLEXES

A catalyst speeds up the rate of a chemical process by providing an alternative mechanism involving a reduced activation energy. The catalyst itself can generally be recovered chemically unchanged after the process (Figure 55).

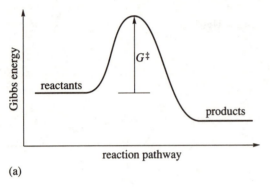
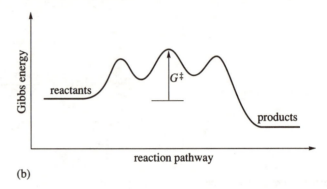

Figure 55 Typical Gibbs free energy profiles for a reaction (a) and for the same reaction employing a catalyst (b). Note that $G^\ddagger$ is lower for the catalysed reaction, even though there are more steps in the profile.

Catalysis is one of the most important aspects of the chemical industry because catalysts speed up chemical change, and so save time and money. In some cases thermodynamically favourable reactions do not occur even on a geological time-scale in the absence of a catalyst.

There are two types of catalysis—**homogeneous** or **heterogeneous**. In heterogeneous catalysis, the catalyst and reactant molecules are in different phases, and the enhanced reactivity occurs at the solid/gas or solid/liquid interface. In homogeneous catalysis the reactants are all in the same phase, normally dissolved in solution.

This last part of our study of organometallic chemistry focuses principally on homogeneous catalysis for a number of reasons.

(a) A number of organic compounds made by industry involve organometallic catalysts at some stage.

(b) It is easier to study the mechanistic details of homogeneous organometallic reactions than it is to study those reactions involving heterogeneous catalysts.

(c) Whole new facets of organic chemistry can occur at a metal centre because coordination to a metal modifies the behaviour of ligands.

Some heterogeneous catalytic systems will be mentioned in comparison to the homogeneous ones. It should be noted that industry uses heterogeneous catalysts wherever possible because separation of the catalyst from the product is much easier. Indeed the majority of current industrial organometallic catalysts are used heterogeneously.

Organometallic catalysts bridge the gap between inorganic and organic chemistry, and combine a number of the principles discussed earlier in the Course.

Requirements for catalysts
Three broad statements can be made about transition-metal catalysis.

- The coordination of reactive ligands to a transition metal brings them close together, thus favouring reaction.

- Through coordination to a transition metal, a particular fragment becomes activated for subsequent reactions (either inter- or intramolecularly).

- Vacant coordination sites must be available at the metal.

To be an ideal catalyst a particular metal should be stable in more than one coordination number and be capable of binding a substrate molecule selectively but not so tightly that it cannot be removed. It should be noted that in this context the complexes of transition metals at the end of the series, the so-called 'platinum' group metals, Ru, Rh, Pd and Pt, make the best catalysts because these elements are able to form complexes that are coordinatively unsaturated.

Two major problems exist with catalysts. The first is that removal or recovery of the catalyst after the reaction is difficult and can be a major cost in a process (especially with expensive metals like platinum); secondly, catalysts are liable to undergo irreversible reactions with trace elements in the substrate medium. For example, traces of sulphur in hydrocarbon feedstocks have to be rigorously removed; otherwise metal–sulphur bonds are formed which poison the catalyst.

R. **Tolman** has articulated a **rule** that seems to govern all transition-metal-mediated catalytic processes: 'detectable concentrations of diamagnetic organometallic complexes of the transition metals exist only if the central metal atom contains sixteen or eighteen electrons'. A consequence of this rule is that organometallic reaction sequences proceed by elementary steps involving intermediates with sixteen or eighteen valence electrons.

8.1 Alkene metathesis

As described in Section 4.3, metathesis is the exchange of atoms between groups. For alkenes, metethasis takes place as follows:

$$CH_2=CR_2 + CH_2=CR_2 \rightleftharpoons \begin{array}{c} CR_2 \\ \| \\ CR_2 \end{array} + \begin{array}{c} CH_2 \\ \| \\ CH_2 \end{array} \quad\quad 94$$

SLC 9

One industrial application of alkene metathesis is the conversion of propene into the more useful ethene (for polymerisation) and pure but-1-ene (to be used in oxidation). Alkene metathesis has less industrial importance than *Ziegler–Natta catalysis*, but is very intriguing from a mechanistic viewpoint because a carbene intermediate has been found to be important in the process. Alkene metathesis can occur homogeneously using a variety of catalysts including $WCl_6/AlClEt_2/EtOH$, or heterogeneously using $[Mo(CO)_6]$ on Al_2O_3.

Figure 56 shows a mechanistic interpretation of the metathesis reaction of alkenes.

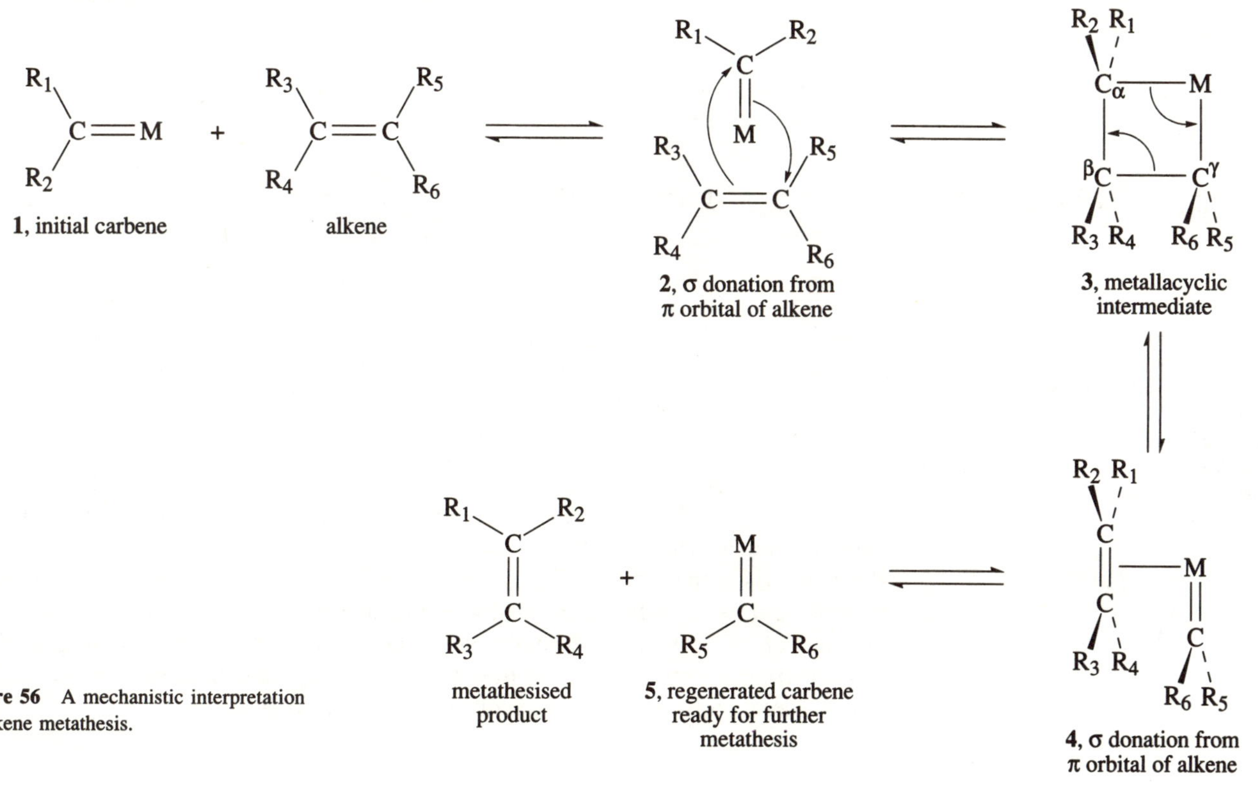

Figure 56 A mechanistic interpretation of alkene metathesis.

The active catalyst in this process is the free carbene in stages **1** and **5**. The intermediate in stage **3** is known as a *metallacycle* because a metal atom is contained in a ring of other atoms; this ring may cleave by breakage of the M—C$_\alpha$ bond, and hence produce the alkene–carbene adduct in stage **4,** or cleave the M—C$_\gamma$ bond, and so revert to the adduct in stage **2**. Only on the pathway towards stage **4** does metathesis occur (that is, an exchange of atoms; in the initial carbene the metal is coordinated to C$_\alpha$, whereas in the final carbene it is coordinated to C$_\gamma$).

Evidence for this process has been obtained by isolating the carbenes and the metallacyclic intermediates, and by using ^{2}H and ^{13}C n.m.r. labelling studies.

8.2 Heterogeneous catalysis: the Ziegler–Natta process

SLC 9

Polythene is a substance that we come into contact with probably everyday in the form of plastic bags and drink containers. The original ICI process for making polythene involved high temperatures and pressures, as mentioned in Block 1. However, in the 1950s Karl Ziegler discovered that, in the presence of TiCl$_4$ and AlEt$_3$ in a hydrocarbon solvent, the polymerisation could be performed at atmospheric pressure and room temperature. Giulio Natta showed that with suitable modification of the catalyst, *syndiotactic polymers* of almost any alkene (CH$_2$=CHR) can be produced. The enormous industrial importance of this subject meant that much of the chemical literature on Ziegler–Natta catalysis is in the form of patents. However, a tremendous body of work has been collected on the mechanism of the catalysed reaction. To date no absolute proof of the mechanism exists (one of the inherent problems of catalysis), but the most likely scheme is outlined in Figure 57.

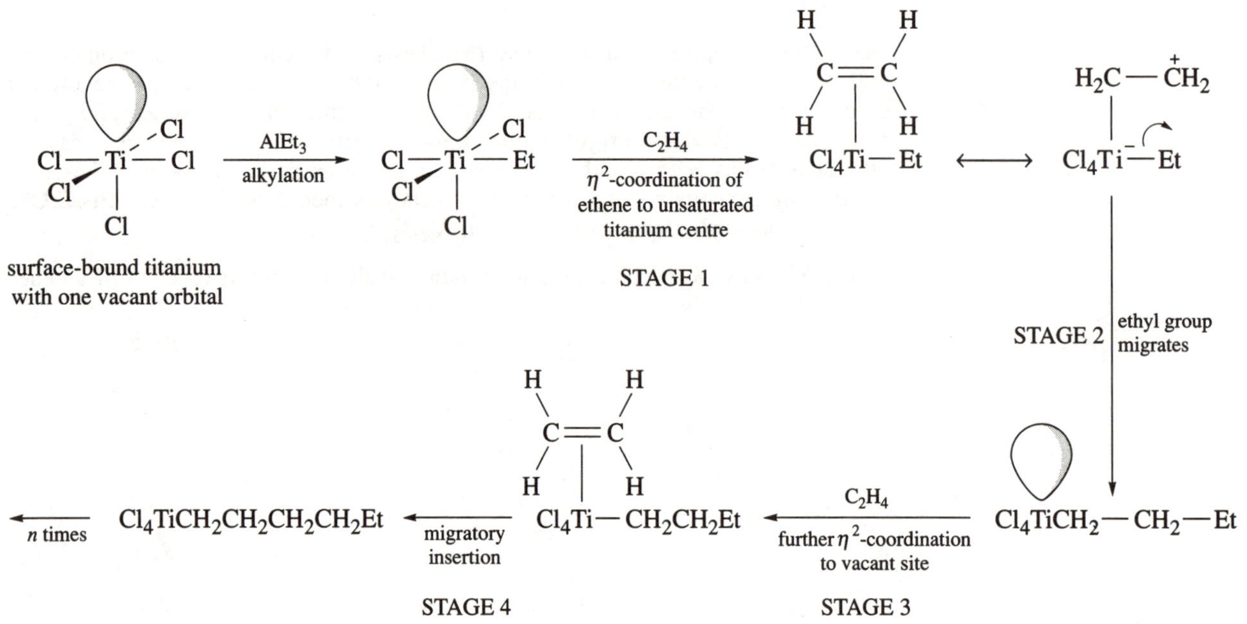

Figure 57 A mechanistic view of Ziegler–Natta polymerisation; the initial structure shows the environment of titanium at the surface of TiCl$_3$. Note that the structures shown in this reaction scheme involve surface-bound titanium; they do not exist as independent molecules (see Block 1, Section 3.7).

A particular advantage of the Ziegler–Natta process is that a more workable polymer with greater strength is produced. The efficiency of the process is indicated by the fact that 5 g of titanium may produce as much as 1 tonne of poly(ethene) or poly(propene).

The reaction is initiated by η^2-coordination of ethane to the partly alkylated Ti^{III} atom (stage 1), followed by the migration of the attached ethyl group from the titanium atom to carbon (stage 2). It is the coordination of the ethene to Ti^{III} which polarises the C=C bond and allows migration of the ethyl group. This concerted process transforms the η^2-bonded ethane ligand into σ-bonded CH$_2$CH$_2$. Coordination of another ethene molecule occurs in stage 3, followed by migratory insertion (ligand combination) of the ethene into the titanium–carbon σ bond in stage 4. These processes occur repeatedly to build up the polymer.

8.3 Homogeneous hydrogenation of alkenes: Wilkinson's catalyst

So far, we have seen sketched examples of alkene metathesis catalysts and Ziegler–Natta heterogeneous catalysts, and we have briefly illustrated how these reactions might occur.

In this Section we shall study in more detail the homogeneous catalytic process that is responsible for hydrogenation of alkenes. The catalytic system is based on the rhodium complex [RhCl(PPh$_3$)$_3$]. The hydrogenation of alkenes is a reaction of major industrial importance. It finds application in the petrochemicals industry, in margarine production (in which many of the components of margarine are prepared by the hydrogenation of liquid vegetable oils), and in the pharmaceutical industry, where the preparation of drugs often involves the hydrogenation of specific double bonds. It is for the hydrogenation of specific double bonds in compounds containing several double bonds that the search for new, more efficient and above all, more selective catalysts has been most vigorous.

One of the most successful hydrogenation catalysts is [RhCl(PPh$_3$)$_3$], which is an effective homogeneous catalyst in solutions of aromatic hydrocarbons such as benzene and toluene. This catalyst was developed by G. Wilkinson of Imperial College, London, and is now almost universally referred to as **Wilkinson's catalyst**. The basic operation of the catalyst is as follows:

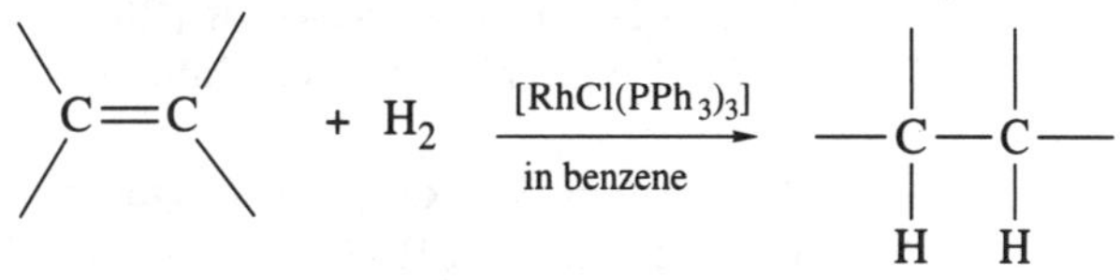

$$\begin{array}{c}\diagdown \\ \diagup\end{array}C=C\begin{array}{c}\diagup \\ \diagdown\end{array} + H_2 \xrightarrow[\text{in benzene}]{[RhCl(PPh_3)_3]} -\underset{H}{\overset{|}{C}}-\underset{H}{\overset{|}{C}}- \qquad 95$$

The ideal hydrogenation catalyst should be capable of catalysing hydrogenations at or near atmospheric pressure, since this eliminates the need for cumbersome and expensive vessels capable of withstanding high pressures. In this respect, the Wilkinson catalyst is remarkably effective for the hydrogenation of isolated double or triple bonds.

In studies of molecular mass, ^{31}P n.m.r. and the electronic spectra of solutions, [RhCl(PPh$_3$)$_3$] has been shown to undergo only minor loss of a triphenylphosphine ligand in solution in tetrahydrofuran. However, when hydrogen is bubbled into the solution, a rapid and reversible reaction occurs:

$$[RhCl(PPh_3)_3] + H_2 \rightleftharpoons [RhCl(H)_2(PPh_3)_3] \qquad 96$$

☐ Why do you think the forward direction of reaction 96 is called an **oxidative addition**? What is a reasonable name for the back reaction?

■ The forward reaction involves oxidation of rhodium(I) to rhodium(III) and addition of hydrogen—hence *oxidative addition*. The back reaction is known as **reductive elimination**.

The product of reaction 96 undergoes reversible loss of a triphenylphosphine ligand to yield a five-coordinate complex. The equilibrium 97 is set up such that a small proportion of the dissociated complex is present at room temperature:

$$[RhCl(H)_2(PPh_3)_3] \rightleftharpoons [RhCl(H)_2(PPh_3)_2] + PPh_3 \qquad 97$$

This five-coordinate complex is said to be *coordinatively unsaturated*, because it is potentially able to pick up a further ligand, as we shall see in a moment.

^{1}H n.m.r studies in chloroform solution show the presence of three peaks in the metal hydride region, which is at lower frequency than TMS (tetramethylsilane). The relative intensity of the peaks is 1 : 0.5 : 0.5 at low field (90 MHz).

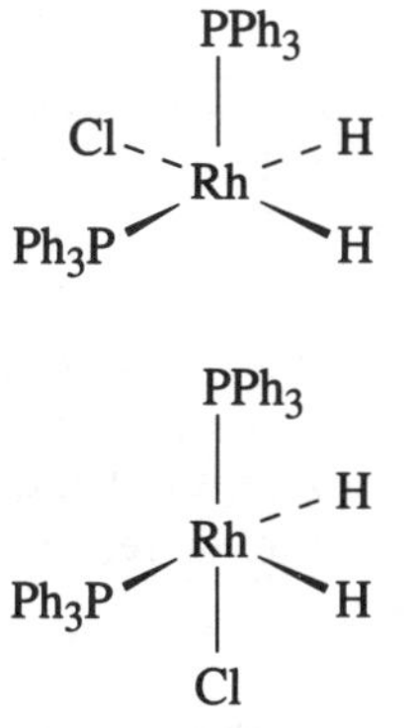

Figure 58 Alternative structures, based on an octahedron with a vacant site, for [RhCl(H)$_2$(PPh$_3$)$_2$], as deduced from ^{1}H n.m.r. spectroscopy.

☐ By reference to the two structures in Figure 58, explain the observed ^{1}H n.m.r. spectrum.

■ For either structure, one hydride resonance would be split into a doublet by the large $^2J(^1H,^{31}P)$ *trans* coupling. The $^2J(^1H,^{31}P)$ *cis* coupling, $^1J(^1H,^{103}Rh)$ coupling and $^2J(^1H,^1H)$ coupling are not resolved well at 90 MHz, but would be seen on a high-field instrument (500 MHz). Also for both structures the resonance for the hydride with no *trans* phosphorus would be seen as a broad singlet. Thus, n.m.r. is unable to distinguish between the two isomers.

The five-coordinate complex $[RhCl(H)_2(PPh_3)_2]$ has a **vacant site**, which may be occupied by an alkene molecule to yield a six-coordinate complex in which the alkene and both hydrogen atoms are bound to the same rhodium atom; this is the complex $[RhCl(H)_2(alkene)(PPh_3)_2]$.

The next stage involves transfer of the hydrogen atoms to the alkene. We can obtain some clues as to how this occurs from a variety of evidence.

☐ Look at each of the following observations, and deduce as much as possible from them yourself before looking at the answers. (Beware that some of the answers may be contradictory.)

1 When a simple straight-chain alkene is hydrogenated with a mixture of H_2 and D_2 in the presence of $[RhCl(PPh_3)_3]$, the major products are alkanes and dideuterioalkanes; very little monodeuterioalkane is formed.

2 When cyclohexene is hydrogenated using a mixture of H_2 and T_2 (tritium) in the presence of $[RhCl(PPh_3)_3]$, some monotritiocyclohexane is formed.

SLC 10

3 Deuteration of *cis*-but-2-endioic acid yields predominantly the *achiral compound* meso-2,3-dideuteriobutanedioic acid (reaction 98):

cis-but-2-endioic acid + D_2 $\xrightarrow{[RhCl(PPh_3)_3]}$ *meso*-2,3-dideuteriobutanedioic acid   98

(You may find molecular models helpful here; the objective is to decide whether or not the reaction is stereoselective, and if it is, in what way.)

4 Partial hydrogenation of hex-2-yne in the presence of Wilkinson's catalyst yields hex-2-ene, in which the *cis* : *trans* ratio is in excess of 20 : 1.

■ 1 This strongly suggests that both H (or D) atoms in H_2 or (D_2) bound to the rhodium are transferred to the same alkene molecule.

2 This is at first sight mutually contradictory with observation 1. However, the conflict can be resolved if the two hydrogen (or deuterium or tritium) atoms are transferred from rhodium to the alkene in two successive steps. The second step generally takes place immediately after the first, but may on occasion be slightly delayed, allowing time for exchange to occur between free hydrogen (or deuterium or tritium) and that bound to rhodium. (For experimental convenience, a mixture of hydrogen and tritium was used by the research workers concerned. However, a similar result can also be observed with hydrogen and deuterium mixtures.)

3 Since *meso*-1,2-dideuteriobutanedioic acid has the stereochemical arrangement shown in reaction 98, it is apparent that both deuterium atoms have added to the same side of the double bond (*cis* addition). Thus, the addition is stereoselective in a *cis* manner.

4   The fact that partial hydrogenation of hex-2-yne to hex-2-ene yields predominantly the *cis* alkene again indicates that the addition is largely stereoselective in a *cis* manner; for example

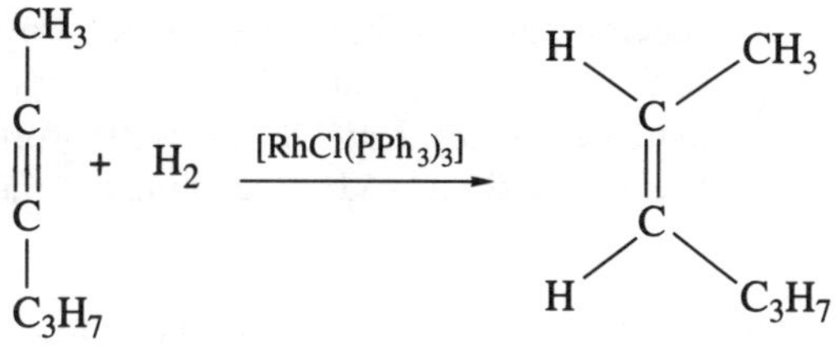

$$\begin{array}{c} CH_3 \\ | \\ C \\ ||| \\ C \\ | \\ C_3H_7 \end{array} + H_2 \xrightarrow{[RhCl(PPh_3)_3]} \begin{array}{c} H \quad CH_3 \\ \diagdown \diagup \\ C \\ || \\ C \\ \diagup \diagdown \\ H \quad C_3H_7 \end{array}$$

99

If both hydrogen atoms were transferred simultaneously to hex-2-yne in observation 4, one would expect only the *cis* isomer to be formed. The small amount of *trans* isomer could be formed in one of two ways, which, without further experiment, cannot be distinguished. These are:

(a)   The two hydrogen atoms are not transferred to the alkyne simultaneously but in rapid succession, giving only limited opportunity for *cis* to *trans* isomerisation. This would be consistent with the deduction made from observations 1 and 2.

(b)   All the hex-2-ene formed by hydrogenation may be the *cis* isomer, but subsequent *cis*-to-*trans* isomerisation may occur.

A mechanism for the hydrogen-atom transfer in [RhCl(H)$_2$(alkene)(PPh$_3$)$_2$] which would be consistent with observations 1 and 3 is shown in reaction 100:

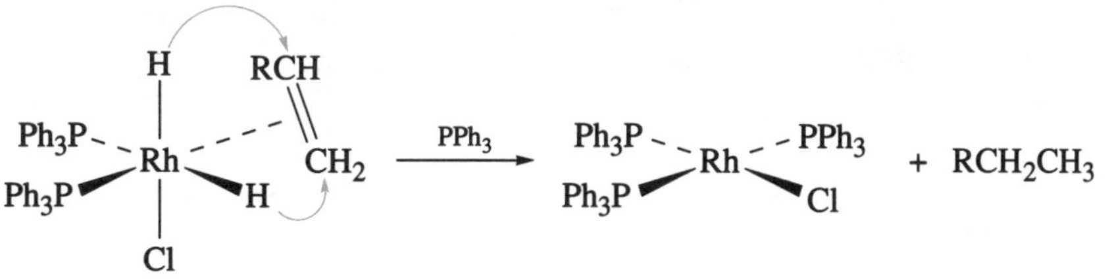

100

(Note that the green curly arrows here and in reaction 101 represent a ligand shift and not electron shifts as the black curly arrows used elsewhere.)

However, this would not be consistent with observation 2. The two-step mechanism shown in reaction 101 would be consistent with observation 2:

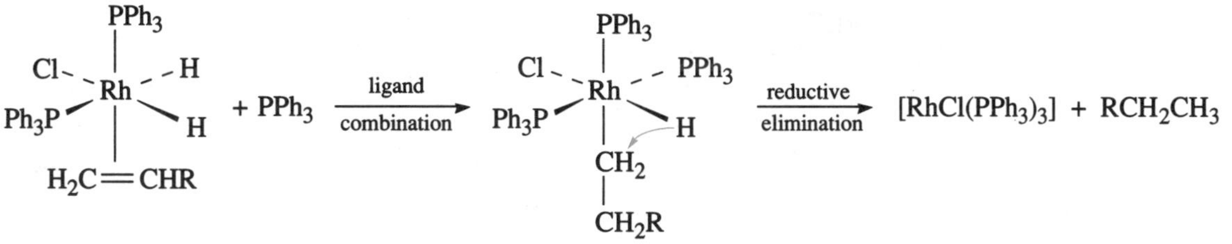

101

In many cases the second step would follow very closely on the first, thereby resulting in no hydrogen exchange, but in some cases there could be sufficient delay between the two steps to allow time for hydrogen exchange at the rhodium(III) to occur. The first step of reaction 101 involves a **ligand combination reaction** of the alkene and one of the hydrogen atoms to yield an alkyl intermediate, and is either accompanied by or followed immediately by, addition of triphenylphosphine to the vacant site created. This is followed by reductive elimination of the alkyl and hydrido groups as alkane, RCH$_2$CH$_3$, and regeneration of [RhCl(PPh$_3$)$_3$], which can then restart the whole process — known generally as a **catalytic cycle**.

A stepwise catalytic mechanism for the hydrogenation of alkenes using Wilkinson's catalyst is shown in Figure 59. Steps 1 + 2 + 3 and 7 + 8 + 9 represent the two alternative pathways for going from [RhCl(PPh$_3$)$_3$] to [RhCl(H)$_2$(RCH=CH$_2$)(PPh$_3$)$_2$].

☐ Examine the cycle in Figure 59. How does it demonstrate that the rhodium complex acts as the catalyst?

■ Careful examination of the cycle shows that the only reactants (that is, the species put into the cycle) are hydrogen and alkene, and the only product (that is, species produced by the cycle) is alkane. Triphenylphosphine is eliminated in stages **2** and **8**, but is reabsorbed in stage **5**. There is no net gain or loss of rhodium, and the complex is therefore a catalyst.

SLC 11 The observations that we have numbered 1, 3 and 4 on p.78 show the *chemoselectivity* and *stereoselectivity* of the process. This is achieved by the Wilkinson catalyst acting as a **template**; that is, it holds the reactants in a unique configuration.

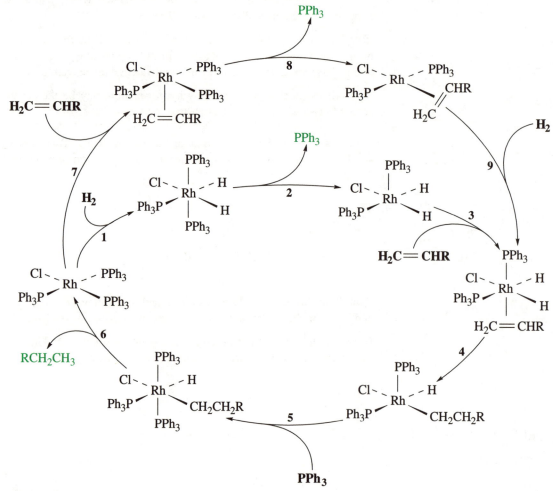

Figure 59 Catalytic cycle for the [RhCl(PPh$_3$)$_3$]-catalysed hydrogenation of alkenes. Note that in a cyclic representation the reaction X + Y ⟶ A is written as

and the reaction B ⟶ W + Z as

B ⟶ W
    ↘ Z

In this and all subsequent catalytic cycles in this Block, reactants are indicated by formulae in bold type and products are shown in green.

SAQ 15 What is the effect on the strength of the alkene double bond when it is coordinated to a metal?

SAQ 16 List the properties that are necessary for a substance to catalyse alkene hydrogenation, and show how the Wilkinson catalyst fulfils each requirement.

SAQ 17 What type of alkenes, if any, might you expect Wilkinson's catalyst to be unable to hydrogenate? Why?

SAQ 18 Give a chemical description for the reactions in steps **1–9** of Figure 59, including the electron count at the metal at each stage. Do the intermediates in this cycle obey Tolman's rule?

□ A study of the kinetics of hydrogenation shows that the addition of alkene to the rhodium complex is the slow step of the reaction. Draw a possible energy profile for the reaction when it follows steps **1–6** of Figure 59.

■ The energy profile would be something like the black curve in Figure 60. The relative values of $G^{\ddagger}$ can be estimated from bond enthalpy data, and the enthalpy changes can be estimated from the relative stabilities of the complexes. The overall value of $\Delta H_m^{\ominus}$ is easily measured. The slow step (**3**) must be rate controlling, so the subsequent steps are indicated by a dashed line. We can see from Figure 60 that the reaction is faster when catalysed because the uncatalysed process, which has a high energy barrier, has been replaced by a multistep process in which all the activation energies are relatively low.

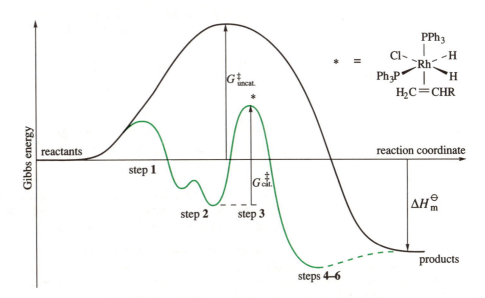

Figure 60 Gibbs free energy profile for hydrogenation of alkenes by Wilkinson's catalyst.

Having looked in some detail at the Wilkinson homogeneous hydrogenation catalyst, it is pertinent to enquire whether it has any applications. The answer is yes, but they are largely specialised applications in the pharmaceutical industry, where the cost of a particular stage of a synthetic process is less important than in many other areas of chemical industry. Within the pharmaceutical industry the Wilkinson catalyst is just one of the hydrogenation catalysts that a manufacturer would consider for carrying out a hydrogenation in a stereoselectively *cis* manner. The principal disadvantages of the Wilkinson catalyst are twofold:

(a) Being a soluble catalyst, it cannot be removed at the end of a reaction by simple filtration in a way that a heterogeneous catalyst can.

(b) It is very expensive to prepare both because of the high cost of rhodium and the high cost of synthesising Wilkinson's catalyst. Although the catalytic cycle showed no loss of rhodium, some of the metal is lost in practice owing to the imperfect separation of alkane and the rhodium complex at the conclusion of the hydrogenation. Separation of the catalyst at the end of the reaction is, in fact, a major problem and is one of the reasons why industry has been unwilling to adopt homogeneous catalysts more widely, preferring instead to use heterogeneous catalysts. One solution to the problem of catalyst separation is to support the homogeneous catalyst on an insoluble support such as a polymer, which allows the catalyst to be separated from products by filtration.

It should be noted that the rate-determining step in the catalytic cycle, the insertion step (step **4** in Figure 59), was only recently discovered by special n.m.r. techniques (J. M. Brown in 1984) and by molecular orbital calculations. Further selectivity in hydrogenation can be achieved by using [RhH(CO)(PPh$_3$)$_2$], which is similar to Wilkinson's catalyst but will hydrogenate only terminal alkenes.

8.4 Fischer–Tropsch catalysis

Fischer–Tropsch catalysis is a catalytic technique that embraces a multitude of catalysts (such as Fe, Fe_3O_4, $[Ru_{12}(CO)_{34}]^{2-}$), reagents and products. Common to all the techniques is the use of simple carbon and hydrogen sources as the raw materials, and the conversion of these into more desirable, longer chain length and more useful hydrocarbon compounds.

Fischer–Tropsch catalysis enables the production of liquid alkanes, methanol, alkenes or higher alcohols. A common feature of a Fischer–Tropsch synthesis is the use of carbon monoxide as the source of carbon:

$$x\text{CO} + y\text{H} \xrightarrow{\text{catalyst}} \text{'CHO' products} \qquad 102$$

TLC 1

The carbon monoxide/hydrogen mixture is known as **synthesis gas** (or *syn gas*). It can be derived from the controlled reaction of petroleum or natural gas with steam, or by coal gasification. At present, natural gas is used as the feedstock:

$$CH_4(g) + H_2O(g) = CO(g) + 3H_2(g); \qquad \Delta H_m^\ominus = +1\,205\,\text{kJ}\,\text{mol}^{-1} \qquad 103$$

syn gas

However, in the future natural gas may be replaced by coal, since the world's coal reserves will far outlast those of natural gas:

$$C(s) + H_2O(g) = CO(g) + H_2(g); \qquad \Delta H_m^\ominus = +132\,\text{kJ}\,\text{mol}^{-1} \qquad 104$$

coal   syn gas

Fischer–Tropsch processes can be heterogeneously or homogeneously catalysed. Both catalytic processes have the problem of selectivity because a variety of products —gasoline, diesel and waxes— can form from the catalytic brew.

Despite the high costs, South Africa has used the Fischer–Tropsch process to produce petrol from coal, because the sale of crude oil to South Africa was banned by trade sanctions. Selectivity can be partly overcome by the use of size-selective secondary catalysts such as zeolites.

Mechanistic models of Fischer–Tropsch reactions can be based on the reaction sequence shown in Figure 61. The **active site** can be on a metal surface (heterogeneous) or a vacant coordination site of a transition-metal complex in solution (homogeneous). The mechanism of the reaction has been probed by synthesising potential organometallic intermediates and by studying reaction patterns.

Figure 61 A mechanistic interpretation of Fischer–Tropsch catalysis. Note that in the schemes shown all the carbon used in the process is derived from carbon monoxide, so that any hydrocarbon synthesised is built up one carbon at a time. Also note the number of competing chemical pathways for the reaction, which means that a number of different products can be formed.

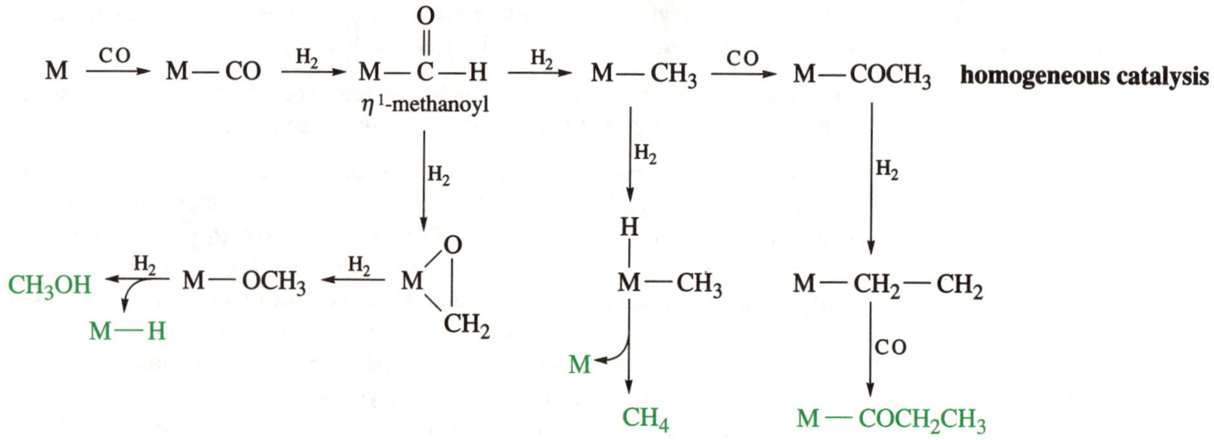

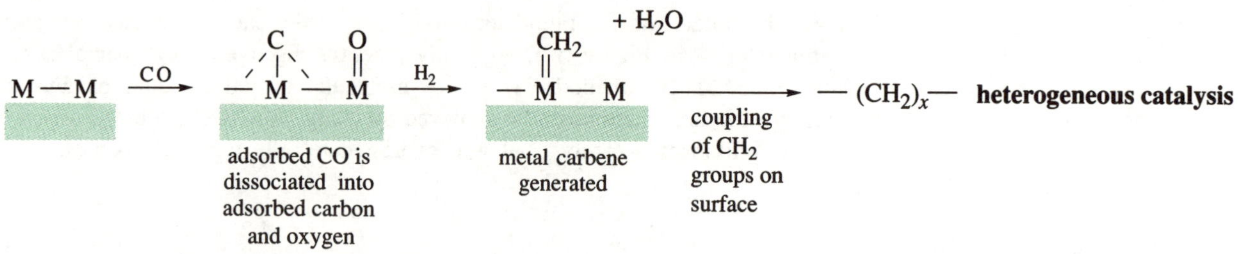

The range of catalysts makes comparative reaction sequences for heterogeneous and homogeneous reactions difficult to rationalise; many schemes have been suggested for Fischer–Tropsch synthesis. However, certain common features can be observed for all processes: synthesis gas provides the source of both carbon and hydrogen, and coupling of CH_2 groups to build up the hydrocarbon chain length is necessary. At present the abundance of oil and gas means that Fischer–Tropsch catalysis using synthesis gas to produce gasoline and waxes has little commercial utility. As the world's resources of oil and natural gas diminish, coal, which has an expected lifetime of about four hundred years, may become the main feedstock for hydrocarbons. In this case more effective Fischer–Tropsch catalysts will be required. It is this prospect that has stimulated much current organometallic research.

8.5 Essential steps in catalytic cycles involving transition-metal complexes

The specific example in Section 8.3, illustrated most of the essential steps or 'building blocks' of homogeneous catalysis. We have also briefly discussed the prerequisites of organometallic catalysts (and Tolman's rule) in Section 8.1 (pp.74–5). In this and the next Section we review the important requirements for catalysis in a little more detail.

8.5.1 Vacant sites

Coordination at a vacant site on a metal complex is a vital component of many catalytic cycles involving alkenes, and has been described as the single most important property of a homogeneous catalyst. This is because alkenes are among the commonest raw materials available to the chemical industry (they can be produced cheaply from crude oil), and they do not form very strong complexes with metals. Accordingly, alkenes would not easily displace other more tightly bound ligands; hence complexes with vacant sites (that is, coordinatively unsaturated complexes) are commonly found to be active catalysts. Vacant sites can arise in at least two distinct ways:

1 In solution by loss of a ligand, as in the reaction

$$[RhCl(H)_2(PPh_3)_3] \rightleftharpoons [RhCl(H)_2(PPh_3)_2] + PPh_3 \qquad 105$$
$$\text{six coordinate} \qquad \text{five coordinate}$$

This dissociation (which you met in Section 8.3) is promoted by the steric bulk of the triphenylphosphine ligands. Dissociation of PPh_3 from $[RhCl(H)_2(PPh_3)_3]$ releases some steric strain in the complex caused by the presence of the three bulky PPh_3 ligands. Clearly there is less repulsion between the two remaining PPh_3 groups in the five-coordinate product.

2 In heterogeneous catalysts, such as $TiCl_3$ in the Ziegler–Natta polymerisation of alkenes, vacant sites arise at the surface of the crystal. In the body of a $TiCl_3$ crystal, each Ti is surrounded octahedrally by six chlorine atoms, each of which bridges to a second Ti atom (structure **6**).

However, some modification of this arrangement is necessary at the surface so as to maintain electrical neutrality and to ensure that the formula is $(TiCl_3)_\infty$ and not $(TiCl_{3-x})_\infty$. It turns out that at the surface of the crystal some Ti atoms are surrounded by only five chlorine atoms, the sixth position being left vacant. Four of the chlorines surrounding such a surface Ti atom link it to other Ti atoms in the crystal and the fifth is surface bound only to a single Ti atom (structure **7**). The sixth coordination position is then a vacant site at the surface of a $TiCl_3$ crystal, which is free to coordinate alkenes and so catalyse their reactions.

8.5.2 Oxidative addition and reductive elimination

An example of the oxidative addition of hydrogen to a metal is shown by reaction 96 (Section 8.3):

$$[RhCl(PPh_3)_3] + H_2 \rightleftharpoons [RhCl(H)_2(PPh_3)_3] \qquad 96$$

There are two ways in which oxidative addition can occur:

$$L_m M^n + A-B \longrightarrow \left[L_m M^{n+II} \begin{matrix} A \\ \diagup \\ \diagdown \\ B \end{matrix} \right] \qquad 106$$

$$L_m M^n + A-B \longrightarrow \left[L_m M^{n+II} - A \right]^+ + B^- \qquad 107$$

(M is the metal, L represents the ligands and A—B indicates, for example, H_2.)

☐ Examine reactions 106 and 107. How has the oxidation state and the coordination number of the metal altered in each reaction?

■ In both reactions, the oxidation state of the metal increases by two. The coordination number also increases, by two in reaction 106 and by one in reaction 107.

Thus, in order to be able to undergo oxidative addition, a metal must have two stable oxidation states separated by two units. As we have seen, the lower oxidation state should normally exhibit a lower coordination number than the higher oxidation state and should preferably be coordinatively unsaturated.

This condition will normally be met, since the lower oxidation state has the larger number of d electrons on the metal ion, and, the maximum coordination number of a metal ion decreases with increase in the number of electrons.

The reverse of reactions 106 and 107 is reductive elimination, in which the oxidation state of the metal decreases and a product is released from the catalyst. Perhaps the most important reductive elimination, because of its involvement in catalytic cycles, is the reductive elimination of alkyl and hydride ligands to yield an alkane:

$$L_m M^{n+II} \begin{matrix} R \\ \diagup \\ \diagdown \\ H \end{matrix} \longrightarrow L_m M^n + RH \qquad 108$$

The final stage of the Wilkinson catalytic cycle involves a reductive elimination (step **6**, Figure 59):

$$\underset{\text{Ph}_3\text{P}}{\overset{\text{PPh}_3}{\underset{|}{\text{Cl}}}} \overset{\text{Cl}}{\underset{\overset{|}{\text{CH}_2}}{\text{Rh}}} \overset{\text{PPh}_3}{\underset{\text{H}}{\diagdown}} \xrightarrow{\text{reductive elimination}} \underset{\text{Ph}_3\text{P}}{\text{Cl}} \text{Rh} \overset{\text{PPh}_3}{\underset{\text{PPh}_3}{\diagdown}} + RCH_2CH_3 \qquad 109$$

Reductive elimination regenerates the catalytic species in its original oxidation state, and thus allows the catalytic cycle to be repeated.

Notice that oxidative addition and reductive elimination very closely parallel the adsorption of molecules on to solid transition-metal surfaces and their subsequent desorption. Other molecules besides hydrogen which can be added reversibly to coordinatively unsaturated metal complexes are haloalkanes and halogens. In addition to alkanes, reductive elimination processes can be used to form aldehydes and acyl chlorides.

8.5.3 Ligand combination reactions

Ligand combination reactions (which are often known as insertion reactions), in which two groups coordinated to the same metal atom combine, generally form the basis of a catalytic process.

☐ Give an example of a ligand combination reaction you have already met.

■ A ligand combination reaction is the vital step in catalytic hydrogenation. It occurs after all the components have been assembled around the metal, when the alkene double bond is weakened by coordination (step **4**, Figure 59):

110

Other important ligand combination reactions involve the combination of alkene and alkyl ligands to give a longer-chain alkyl group. Such a reaction forms the heart of the Ziegler–Natta polymerisation of alkenes (Section 8.2).

SAQ 19 (a) What are the key requirements for homogeneous catalysis in solution?

(b) What are the main differences between heterogeneous and homogeneous catalysis?

SAQ 20 The Monsanto ethanoic acid process is the homogeneously catalysed carbonylation of methanol, which yields $c.\ 10^6$ tonnes/annum of ethanoic acid. The mechanism for the reaction has been formulated as shown in Figure 62.

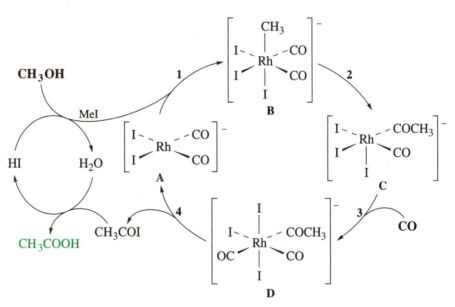

Figure 62 The Monsanto ethanoic acid process. This process involves the carbonylation of methanol to ethanoic acid using $[RhI_2(CO)_2]^-$ as a catalyst.

(a) What is the oxidation state of rhodium in the organometallic intermediates **A**, **B**, **C** and **D** in this catalytic cycle?

(b) Does Tolman's sixteen/eighteen-electron rule apply to the intermediates **A**, **B**, **C** and **D** in this cycle?

(c) What are the names for steps **1**, **2**, **3** and **4**?

(d) How many separate cycles operate in the above scheme?

(e) Which compounds are coordinatively unsaturated in the cycle?

SAQ 21 The process shown in Figure 63 represents the Union Carbide hydroformlyation process for the production of aldehydes.

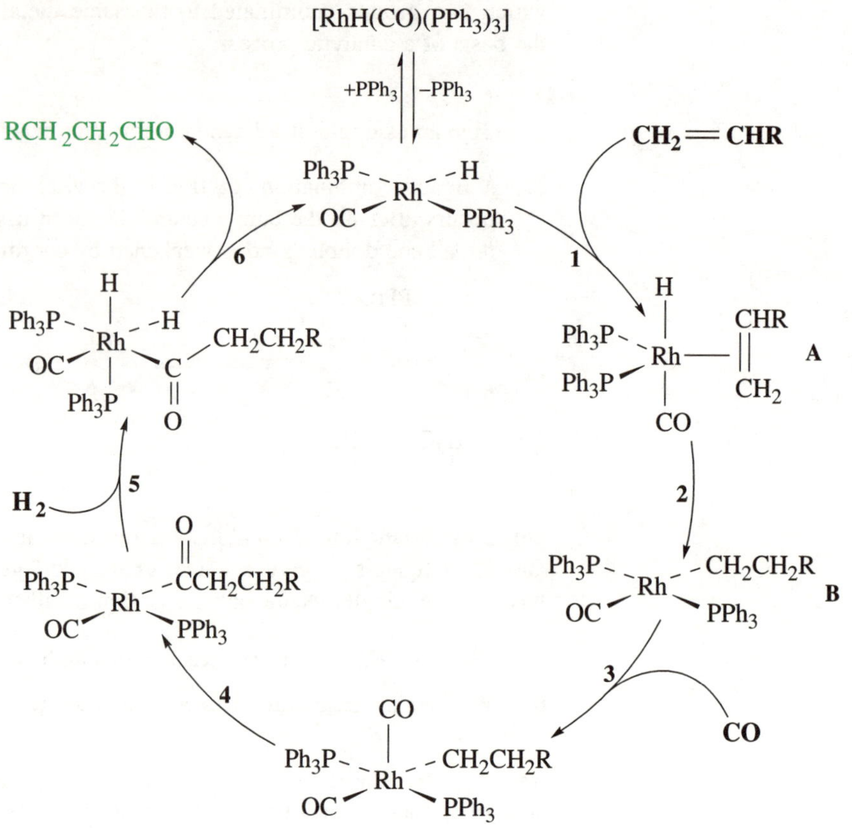

Figure 63 The Union Carbide hydroformylation process. This process converts alkenes into aldehydes using [RhH(CO)(PPh$_3$)$_3$] as a catalyst.

(a) Give descriptions for steps **1–6**, including any reaction names.

(b) Do compounds **A** and **B** obey the eighteen-electron rule?

(c) Describe the metal–carbon bonding in compound **A**.

SAQ 22 The catalytic cycle shown in Figure 64 represents the hydrocyanation reaction, which is important for the commercial production of hexane-1,6-dinitrile from butadiene.

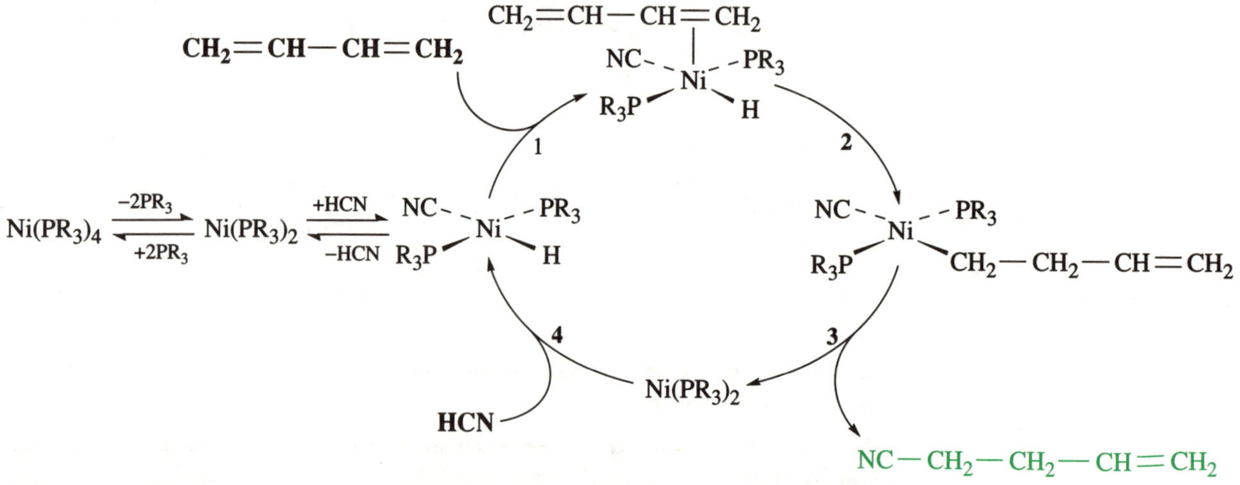

Figure 64 The hydrocyanation reaction, for conversion of alkenes into nitriles using a nickel-based catalyst.

(a) Describe the chemical reactions occurring in steps **1, 2, 3** and **4**.

(b) How is the active catalyst [Ni(CN)HL$_2$] initially formed? (L could be a variety of ligands, but is shown here as PR$_3$.)

(c) Do any of the proposed catalytic intermediates have unusual electron counts?

8.5.4 Attack on a coordinated ligand

Very closely related to ligand combination reactions are reactions in which a coordinated ligand is attacked by a reagent that is not coordinated to the metal. Important examples are provided by attempts to use complexes involving the dinitrogen ligand, N_2, as a means of rendering nitrogen susceptible to attack by hydrogen to yield reduced products such as hydrazine or ammonia:

$$M-N_2 + \text{'H'} \longrightarrow \begin{array}{c} N_2H_4 \\ \text{or} \\ NH_3 \end{array} \qquad 111$$

As you will see in Block 7, the final goal of *nitrogen fixation* is still some way off. However, one reaction that goes some way towards it is that between hydrochloric acid and a *bis*(dinitrogen)tungsten complex, in which one N_2 is eliminated and the other is reduced to an N_2H_4 ligand.

8.5.5 β-Hydrogen abstraction

A reaction that you met on several occasions earlier in this Block, and which bedevils many catalytic processes involving alkene substrates, is β-hydrogen abstraction from a metal alkyl. The reaction may be written as:

$$\begin{array}{c} M-CH_2 \\ | \\ H-CH \\ | \\ R \end{array} \longrightarrow \begin{array}{c} M \\ | \\ H \end{array} + \begin{array}{c} CH_2 \\ \| \\ CH \\ | \\ R \end{array} \qquad 112$$

The reason that this reaction causes problems is that it provides a perfect route for alkene isomerisation (Figure 65).

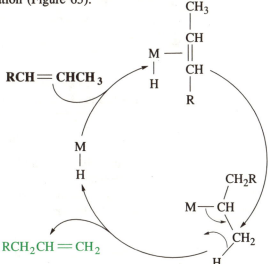

Figure 65 Alkene isomerisation involving β-hydrogen abstraction.

In the course of the hydrogenation of a monoalkene, any alkene isomerisation cannot be detected when the final alkene is isolated, but in the hydroformylation mentioned in SAQ 21 it can lead to a mixture of terminal and internal aldehydes (and hence alcohols following hydrogenation) being obtained as the final products (Figure 66). Clearly a manufacturer wants to produce as pure a product as possible, since for many applications, mixtures of isomers will be unacceptable. Separation generally requires fairly efficient and hence expensive fractional distillation.

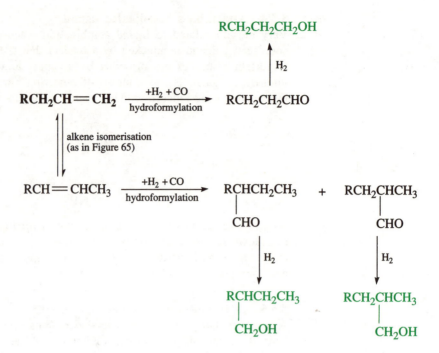

Figure 66 The formation of both terminal and internal aldehydes (and alcohols) by hydroformylation as a consequence of alkene isomerisation.

8.6 Properties of metals in homogeneous catalysis

We can now collect together those properties of the metal that are essential for it to carry out the catalytic processes that we have been studying.

☐ Write down as many of these properties as you can think of, and then compare your list to the one below.

■ 1 At least two stable oxidation states (separated by two units) should exist.

2 It should be capable of forming complexes with a range of coordination numbers, and of forming moderately stable coordinatively unsaturated species.

3 It should be capable of forming a hydride (either a monohydride or dihydride) by direct reaction with gaseous hydrogen.

4 It must have the ability to match the substrate orbitals; if the substrate is an alkene, this means that the metal must have an empty orbital capable of accepting σ donation from the alkene and a filled orbital capable of π back-donation to the alkene (see Figure 27).

Although there is quite a wide range of metals that individually can meet each of the requirements listed above, you will find that the 'platinum group metals', Fe, Ru, Os, Co, Rh, Ir, Ni, Pd and Pt, will meet all of them most effectively. Unfortunately these elements are extremely expensive except for the first-row metals Fe, Co and Ni. The most versatile catalysts (that is, those capable of catalysing the widest range of reactions) are also platinum group complexes. However, other metal ions throughout the Periodic Table catalyse specific reactions where they are able to meet the more limited requirements of one particular reaction.

It is somewhat artificial to separate the properties of the metal and the ligand, since these are closely interrelated. This is particularly true when the ligand *trans* to the site at which the substrate coordinates is considered. Thus, the electronic properties of a given ligand most affect the site *trans* to that ligand, since mutually *trans* sites have more metal orbitals in common than any other two sites. However, the transmission of electronic effects between two mutually *trans* sites is dependent not only on the identity of the metal but also on the oxidation state of the metal. Furthermore, the influence of a ligand on its *trans* site will involve not only an influence on the coordination of a group at the *trans* site, but also an influence on the substitution of that group or other reactions involving it.

8.7 Properties of ligands in homogeneous catalysis

The choice of the ligands that should be present in a catalyst depends on several properties.

8.7.1 Electronic properties of the ligand

Since the catalyst has to coordinate the substrate, it is essential, as mentioned above, that the energy levels of the metal orbitals match as closely as possible the energies of those required by the substrate. 'Coarse tuning' can be effected by balancing the σ donor/π acceptor abilities of the ligands surrounding the metal. Replacement of a powerful σ donor by a weak σ donor that is also a π acceptor will markedly reduce the electron density on the metal ion, and make it more susceptible to nucleophilic attack.

☐ List a range of ligands that are (a) σ donors but not π acceptors, and (b) are both σ donors and π acceptors.

■ (a) σ donors but not π acceptors: hydride, halides, saturated oxygen and nitrogen donors, alkyl.

(b) σ donors that are also π acceptors: carbon monoxide, dinitrogen (N_2), cyanide (CN^-), alkenes, alkynes, tertiary phosphines.

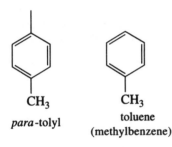

para-tolyl

toluene (methylbenzene)

'Fine tuning' of the electronic properties of the metal is also very important. Replacing Cl^- by Br^-, or P(alkyl)$_3$ by P(aryl)$_3$ can achieve this. Thus, [RhBr(PPh$_3$)$_3$] and [RhCl{P(*p*-tolyl)$_3$}$_3$], are better hydrogenation catalysts for monoalkenes than [RhCl(PPh$_3$)$_3$], whereas [RhCl(PMe$_3$)$_3$] is a poorer catalyst.

8.7.2 Charge on the ligand

Very closely related to the electronic properties is the charge on the ligands. For example, cyanide ions and carbon monoxide are isoelectronic, and thus might be expected to behave similarly. However, because of the charge on the cyanide, they do not. Either the complex carries a negative charge, in which case it is more susceptible to electrophilic than nucleophilic attack, or the cyanide forms complexes with metal ions in higher oxidation states.

8.7.3 Steric effects

The sheer bulk of the ligands can be important in restricting the access of reagents, particularly large ones, to the active site. Thus, for example, the rate of hydrogenation of hex-1-ene decreases by a factor of 775 if the sterically less-hindered phosphine P(*p*-tolyl)$_3$ in [RhCl(phosphine)$_3$] is replaced by the more-hindered *ortho*-tolylphosphine, P(*o*-tolyl)$_3$.

ortho-tolyl

8.8 Summary of Section 8

1 A catalyst speeds up the rate of a chemical process by providing an alternative mechanism involving a reduced activation energy. Catalysis can be either homogeneous or heterogeneous.

2 Homogeneous catalysis using transition metals requires (a) vacant coordination sites at the metal, (b) activation of the ligand on coordination, (c) coordination of the reactive partners on the transition metal so as to favour reaction.

3 Tolman's rule states that detectable concentrations of diamagnetic organometallic complexes of transition metals exist only if the central metal atom contains sixteen or eighteen electrons. Consequently, organometallic reaction sequences proceed by elementary steps, involving intermediates having sixteen or eighteen valence electrons.

4 Ziegler–Natta catalysis is important for the production of processable polymers, the key reaction in the process being the insertion of an alkene into a titanium–alkyl bond.

5 Wilkinson's catalyst [RhCl(PPh$_3$)$_3$] is an effective homogeneous catalyst for the hydrogenation of double bonds. This process can be represented by a catalytic cycle in which a mechanism is postulated for the transformation whereby all the intermediates are represented.

6 Key features of a catalytic cycle can include oxidative addition, reductive elimination, ligand combination (insertion reaction), addition, dissociation. This means that for homogeneous transition-metal catalysts, vacant sites should exist on the metal. Normally at least two stable oxidation states should be available to the metal.

7 In homogeneous catalysis the properties of the metal and ligands are both crucial. The electronic properties of the ligand (σ donor, π acceptor), the size and extent of steric hindrance of the ligands, and the charge on the ligands are vital in catalysis.

8 Fischer–Tropsch catalysis can be used to synthesise a range of hydrocarbons and oxygenated hydrocarbons by using synthesis gas (a mixture of carbon monoxide and hydrogen) as a feedstock.

9 SUMMARY OF BLOCK 6

Today, there are few metals that are not known to have an extensive organometallic chemistry. In this Block, there has been space only to give you a feel for some of the basic compound types and their reactions; there are many aspects which we have had to leave unexplored. In recent years, an extensive organometallic chemistry of the lanthanides and actinides has evolved, although special problems such as air sensitivity and lability has limited research. Organometallic reagents have found special favour with synthetic organic chemists, who now have the power to perform many, previously difficult, transformations cleanly and under mild conditions. We are now able to predict the structure of many organometallic clusters, and predict the site of most likely reactivity within these complexes. In industry, organometallic complexes are of particular value in the search for new catalysts.

Apart from major applications of organometallic compounds in other areas, the variety of structure and bonding which they exhibit is sufficient justification for their study. Many new ideas have developed just from research in this area.

OBJECTIVES FOR BLOCK 6

Now that you have completed Block 6, you should be able to do the following things:

1 Recognise valid definitions of, and use in a correct context, the terms, concepts and principles in Table A.

Table A List of scientific terms, concepts and principles used in Block 6

Term	Page No.
α-hydrogen elimination	14
active site	82
agostic interaction	38
anti-Markownikoff addition	25
arachno structure	56

Term	Page No.
β-elimination reaction	14
back-bonding (back-donation)	41
Buckminsterfullerene (Bucky balls)	47
carbene	24
catalytic cycle	79
closo structure	54
coordinative unsaturation	33
double bond rule	26
electron-count deficiency	50
1 100–1 000 cm^{-1} rule	71
Fischer carbene	35
Fischer–Tropsch reaction	82
four-centre bond	11
Green's rules	61
hapticity	6
heterogeneous catalysis	74
highest-occupied molecular orbital	61
homogeneous catalysis	74
homoleptic ring	29
hydroboration	24
hydrometallation	17
hypo structure	56
insertion reaction	32
isolobal analogy	50
lability	15
ligand combination reaction	79
metallation	17
metathesis	17
molecular organic chemical vapour deposition (MOCVD)	30
nido structure	56
orthometallation	14
oxidative addition	77
reductive elimination	77
ring-whizzing	71
sandwich compound	64
Schrock carbene	35
silicone	27
structural unit	55
substrate	21
synergic process	41
synthesis gas	82
template	80
Tolman's rule	75
transmetallation	17
vacant site	78
Wade–Mingos rules	54
Wilkinson's catalyst	77

2 Given sufficient information, classify the organic ligands in organometallic compounds as:

(a) monohapto, dihapto or polyhapto;

(b) σ, π- or μ-bonded;

(c) even, odd, open or closed. (SAQs 1 and 8)

3 Predict the type(s) of organometallic compound likely to be formed (a) by a particular metal by considering its position in the Periodic Table, and (b) by a particular organic group or molecule. (SAQ 1)

4 Relate the physical and chemical properties of organometallic compounds to their structures and vice versa. (SAQs 2 and 3)

5 Relate the structures of compounds and complex ions containing *polyhapto* ligands to the eighteen-electron rule, and rationalise the complexes that occur in catalytic cycles in terms of Tolman's rule. (SAQs 9, 18, 20, 21 and 22)

6 Give a bonding interpretation of the structures of organometallic species. (SAQ 7)

7 Recognise, and give specific examples of, the various bonding modes (*hapticities*) possible for a particular ligand such as an alkyl or cyclopentadienyl group. (SAQs 13 and 14)

8 Cite evidence to show that the stability and reactivity of organometallic compounds containing *monohapto* ligands reflect:

(a) the strength and polarity of their M—C bonds;

(b) the Lewis acid or Lewis base character of the compound;

(c) the electronic and steric effects of the groups attached to M and C. (SAQ 2)

9 (a) Identify features leading to ready fragmentation of transition-metal complexes containing *monohapto* ligands, and describe the factors that lead to an enhancement of their stability.

(b) Identify factors contributing to the stability or reactivity of a particular organometallic compound, and predict stability or reactivity sequences for related series of organometallic compounds. (SAQs 4 and 5)

10 (a) Identify the types of compound that require special conditions, such as the exclusion of air and water, for their synthesis.

(b) Describe the reactions that require these special conditions. (SAQ 3)

11 Suggest appropriate reagents for the synthesis of selected organometallic compounds. (SAQs 3, 5 and 14)

12 Recognise similarities and differences between organometallic and related organic and inorganic compounds. (SAQ 2).

13 Use Green's rules to predict the site of nucleophilic attack on cationic polyene systems. (SAQ 12)

14 Be able to apply the isolobal analogy to determine isolobal transition-metal and main-Group molecular fragments. (SAQ 10)

15 Use the Wade–Mingos rules to predict the shape of organometallic clusters. (SAQ 11)

16 Identify Fischer- and Schrock-type carbenes, explain their reactivity with electrophiles and nucleophiles, and describe their role in metathesis reactions. (SAQ 6)

17 Explain the agostic interaction and predict the spectroscopic properties of an agostic hydrogen.

18 Identify the factors that can stabilise R—E=E—R and C=E bonds, where E = P, As, Sb, Bi, Se and Te.

19 Identify the types of reaction important in heterogeneous and homogeneous catalysis. (SAQ 19)

20 Give examples of major applications of organometallic compounds.

21 Write down a catalytic cycle for an alkene reaction on a transition-metal complex, and show the importance of:

(a) vacant sites on the complex;

(b) oxidative addition of reagent(s);

(c) ligand combination reactions (insertion reactions);

(d) attack on a coordinated ligand by a reagent;

(e) reductive elimination. (SAQs 18, 20, 21 and 22)

22 Give examples of reactions to show how the reactivity of a π-bonded alkene ligand depends on factors (a)–(d) below, or design suitable complex catalysts with these factors in mind:

(a) the nature of the metal, in particular the range (and stability) of coordination numbers and oxidation states of which it is capable;

(b) the nature of the metal–alkene bond (for example, the susceptibility of the alkene to nucleophilic attack may be increased by loss of electron density on the metal);

(c) the electronic and steric effects of other groups on the metal;

(d) the electronic and steric effects of alkene substituents. (SAQs 15 and 16)

23 Give (or recognise examples of) template or stereoselective effects in an alkene reaction catalysed by a transition-metal complex. (SAQs 17 and 18)

24 Devise (or comment critically on) experiments that test mechanisms of homogeneous catalysis of an alkene reaction by a transition-metal complex. (SAQs 20, 21, and 22)

SAQ ANSWERS AND COMMENTS

SAQ 1
(Objectives 1, 2 and 3)
SLC 12

The way in which the metals are grouped in Figure 5 is reminiscent of the Second Level Course treatment of the *classification of metal hydrides* into (a) ionic, (b) hydrogen bridged, (c) volatile covalent and (d) metal-like or interstitial. This is because both classifications, whether of organo or hydride derivatives, reflect the various bonding capabilities (degree of electropositive character, numbers of orbitals and electrons the atoms can provide for bonding) of the different metals.

SAQ 2
(Objectives 4, 8 and 12)

$Si(CH_3)_4$ is inert to hydrolysis, but the formally isoelectronic $SiCl_4$ reacts very vigorously with water. $SiCl_4$ also functions as a strong Lewis acid, whereas $Si(CH_3)_4$ does not. One reason for these differences is the greater polarity of silicon–chlorine bonds compared with silicon–carbon bonds: the silicon atom of $SiCl_4$ carries a greater positive charge than the silicon atom of $Si(CH_3)_4$, and so is more susceptible to nucleophilic attack. The greater the positive charge on silicon, the more contracted will its lowest-energy vacant orbitals (3d) become, and the greater the facility for bonding to Lewis bases. Another reason why $Si(CH_3)_4$ is less readily hydrolysed is steric hindrance. Methyl groups are bigger than chlorine atoms, so even if there were little electronic difference between $SiCl_4$ and $Si(CH_3)_4$, the latter would be expected to be less susceptible to hydrolysis because its silicon atoms are less accessible to approaching water molecules.

SAQ 3
(Objectives 10 and 11)

(a) Silicon halide ($SiCl_4$ preferred), plus a suitable proportion of a reactive metal methyl, for example LiMe, MeMgX, $AlMe_3$, $ZnMe_2$, $HgMe_2$, $PbMe_4$. This route is illustrated by reaction 113:

$$SiCl_4 + 4MeMgCl \longrightarrow SiMe_4 + 4MgCl_2 \qquad 113$$

(b) $$EtCl + Mg \xrightarrow{ether} EtMgCl \qquad 114$$

(c) Two routes are available: (i) the hydroboration reaction between C_2H_4 and BH_3/thf or $BH_3.OR_2$ in a suitable ether (R_2O) solvent: (ii) the reaction between a boron halide (BCl_3 or BF_3, preferably as $BF_3.OEt_2$) and a reactive metal ethyl, for example LiEt, EtMgBr, $AlEt_3$, $ZnEt_2$, $HgEt_2$, $PbEt_4$:

$$BCl_3 + 3LiEt \longrightarrow BEt_3 + 3LiCl \qquad 115$$

(d) Phenyl magnesium halide plus thallium trichloride:

$$2PhMgX + TlCl_3 \longrightarrow TlPh_2Cl + 2MgXCl \qquad 116$$

LiPh + $TlCl_3$ tends to give $TlPh_3$.

(e) Tin(IV) halide plus a suitable proportion of a reactive metal ethyl (see part (a)); for example

$$SnCl_4 + 2HgEt_2 \longrightarrow SnEt_4 + 2HgCl_2 \qquad 117$$

SAQ 4
(Objective 9)

The alkyl groups most likely to form stable transition-metal complexes are: CH_3 and $CH_2C_6H_5$, since these groups contain no β-hydrogens. CH_2SiHMe_2 forms relatively stable complexes because silicon does not readily form a simple silicon–carbon multiple bond.

SAQ 5
(Objectives 9 and 11)

The important factors are firstly the absence of a β-CH group, which prevents ready elimination of an alkene, and secondly the intermediate formation of a hydrogen–alkene complex prior to loss of the alkene is not possible because there is only one carbon atom in a methyl group. Also, α-hydrogen elimination to generate an alkene is unlikely because tungsten cannot be oxidised above W^{VI}, although the tungsten can accommodate a higher coordination number (cf. the seven-coordinate tungsten complexes discussed in the Laboratory Techniques video programme).

Hexamethyltungsten can be prepared by the following route:

$$WCl_6 + 6LiCH_3 \longrightarrow [W(CH_3)_6] + 6LiCl \qquad 118$$

Other methyl derivatives of main-Group elements, such as MeMgCl, are also suitable for the preparation.

SAQ 6
(Objective 16)

In complexes **d, e, f** and **g** the metal is in oxidation state +5; in complexes **a, b** and **c** the metal is in the +2 oxidation state; in complex **h** it is in oxidation state +4.

Complexes **d, e, f** and **g** are all Schrock-type carbenes because the metal is in a high oxidation state and contains weak π-acceptor ligands. In these complexes the alkylidene carbon atom is nucleophilic and is thus susceptible to electrophilic attack.

Complexes **a, b, c** and **h** are Fischer-type carbenes, because the metal is in a low oxidation state and because of the presence of strong π-acceptor ligands such as CO. In these complexes the carbenes are electron deficient because π electron density is withdrawn by the CO from the metal.

Note the unusual situation of high oxidation state metal complexes having nucleophilic carbene ligands.

SAQ 7
(Objective 6)

The combination of atomic orbitals for the allyl group, together with the metal d orbitals of suitable symmetry for bonding are shown in Figure 67. The metal is placed below the plane of the η^3-allyl group, approximately equidistant from the three carbon atoms, and since each carbon atom provides one electron to the allyl π system, it contributes three electrons to the molecular orbitals of the complex. The π_b orbital is involved in σ-bonding to the metal, whereas π_n and π_a are involved in π- and δ-bonding, respectively.

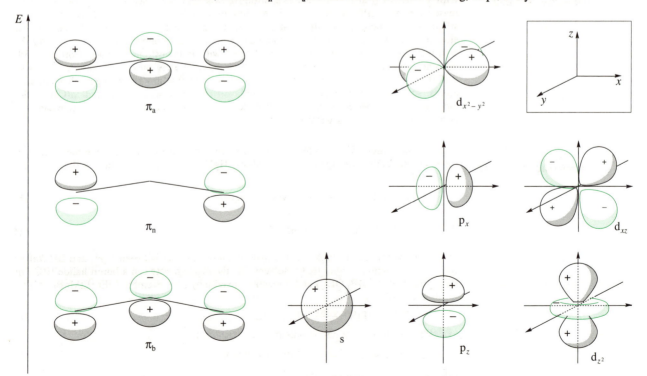

Figure 67 The orbitals of the allyl group with metal atomic orbitals having appropriate symmetry and energy for bonding.

SAQ 8
(Objective 2)

I The bridging cyclo-octatetraene molecule acts as a *tetrahapto* ligand to both the iron atoms. Hence it is μ-bonded. Because the alkene π system is involved in bonding to the metal and the carbon–carbon multiple bonding is retained to some extent, the organo ligand may be regarded as π-bound.

II The central carbon–carbon bond of the $C_2(CN)_4$ ligands is represented in the diagram as a carbon–carbon single bond (*c.* 154 pm). Consequently, the *dihapto* ligand is best considered to be non-bridged and σ,σ-bonded (two metal–carbon σ bonds) to the iridium.

III The organo ligand forms two σ bonds to one iron atom (that is, it is *dihapto*), producing a five-membered ring incorporating iron. The organo ligand also acts as a non-bridged *tetrahapto* ligand to the second iron atom. Multiple bonding is retained to some extent in the organo group, which may therefore be classed as π-bonded to the iron atom beneath the ring.

IV The alkyne ligand is bridging the two cobalt atoms and thus is μ-bonded. The ligand donates from both carbon atoms and thus is *dihapto*. Multiple bonding is retained by the alkyne and so it may be classified as π-bonded to both metal atoms.

V The carbene group :C(Ph)OMe is a *monohapto*, non-bridging ligand, which provides two electrons for the metal–carbene bond. Since the interaction of the carbene with the metal does somewhat involve the carbene π system, the complex is regarded as a σ complex, with a π component.

SAQ 9
(Objective 5)

(a) $ReCl(CO)_4$ appears to be a sixteen-electron system (7 (Re) + 1 (Cl) + 4 × 2 (CO)), but an eighteen-electron system is obtained if the chlorines occupy bridging positions, providing a total of three electrons each to the two rhenium atoms. Hence the structure is best represented by the molecular formula $[Re_2(\mu\text{-}Cl)_2(CO)_8]$ (see Figure 68a).

(b) $FeI(CO)_4$ appears to be a seventeen-electron system (8 (Fe) + 1 (I) + 4 × 2 (CO)), but the unit dimerises by formation of an iron–iron bond, giving an eighteen-electron system (Figure 68b). Thus, the additional electron required is provided by the second iron atom.

(c) $MnCl(CO)_6$ appears to be a twenty-electron system (7 (Mn) + 1 (Cl) + 6 × 2 (CO)); but an eighteen-electron system is obtained if the chlorine electrons are not available to the metal and the complex is ionised (Figure 68c). Thus, the complex will be of the form $[Mn(CO)_6]^+ Cl^-$ (that is, 7 (Mn) + 6 × 2 (CO) − 1 (positive charge)).

Figure 68 Structures of: (a) $Re(CO)_4Cl$, (b) $FeI(CO)_4$ and (c) $MnCl(CO)_6$.

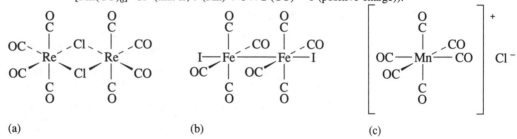

SAQ 10
(Objective 14)

(a) $Mn(CO)_5$ is an example of a d^7 ML_5 fragment. The total number of electrons is 7 + 10 = 17; the electron-count deficiency is therefore 18 − 17 = 1, and the number of half-filled orbitals is 1.

·CH_3. The total number of electrons is 4 + 3 = 7; the electron-count deficiency is therefore 8 − 7 = 1 (8 electrons are required to fill the valence shell of a non-metal), and the number of half-filled orbitals is 1.

An As atom has 5 valence electrons; the electron-count deficiency is therefore 8 − 5 = 3, and the number of half-filled orbitals is 3.

Therefore

$$Mn(CO)_5 \longleftrightarrow \cdot CH_3 \longleftrightarrow\!\!\!\!\!/\,\, As$$

(b) ·CH has 5 valence electrons; the electron-count deficiency is therefore 8 − 5 = 3, and the number of half-filled orbitals is 3.

Phosphorus has 5 valence electrons; the electron-count deficiency is therefore 8 − 5 = 3, and the number of half-filled orbitals is 3.

BH^- has 5 valence electrons (3 + 1 + 1); the electron-count deficiency is therefore 8 − 5 = 3, and the number of half-filled orbitals is 3.

S^+ has 5 valence electrons; the electron-count deficiency is therefore 8 − 5 = 3, and the number of half-filled orbitals is 3.

Se has 6 valence electrons; the electron-count deficiency is therefore 8 − 6 = 2, and the number of half-filled orbitals is 2.

Therefore

$$\cdot CH \longleftrightarrow P \longleftrightarrow BH^- \longleftrightarrow S^+ \longleftrightarrow\!\!\!\!\!/\,\, Se$$

(c) $\cdot\overset{\cdot}{C}H$ has 5 valence electrons; the electron-count deficiency is therefore $8 - 5 = 3$, and the number of half-filled orbitals is 3 (and one C—H bond).

Arsenic has 5 valence electrons; the electron-count deficiency is therefore $8 - 5 = 3$, and the number of half-filled orbitals is 3 (and one lone pair).

$Co(CO)_3$ is an example of a d^9 ML_3 fragment. The total number of electrons is $9 + 6 = 15$; the electron-count deficiency is therefore $18 - 15 = 3$, and the number of half-filled orbitals is 3.

Sn^- has 5 valence electrons; the electron-count deficiency is therefore $8 - 5 = 3$, and the number of half-filled orbitals is 3.

The four fragments are therefore isolobal:

$$\cdot\overset{\cdot}{C}H \longleftrightarrow As \longleftrightarrow Co(CO)_3 \longleftrightarrow Sn^-$$

(d) $[Re(CO)_4]^-$ is an example of a d^7 ML_4 fragment. The total number of electrons is $7 + 8 + 1 = 16$; the electron-count deficiency is therefore $18 - 16 = 2$, and the number of half-filled orbitals is 2.

$Os(CO)_4$ is an example of a d^8 ML_4 fragment. The total number of electrons is $8 + 8 = 16$; the electron-count deficiency is therefore $18 - 16 = 2$, and the number of half-filled orbitals is 2.

$\cdot\overset{\cdot}{C}H_2$ has six valence electrons; the electron-count deficiency is therefore $8 - 6 = 2$, and the number of half-filled orbitals is 2.

The three fragments are therefore isolobal:

$$[Re(CO)_4]^- \longleftrightarrow Os(CO)_4 \longleftrightarrow \cdot\overset{\cdot}{C}H_2$$

(e) $Ir(PR_3)_4$ is an example of a d^9 ML_4 fragment. The total number of electrons is $9 + 8 = 17$; the electron-count deficiency is therefore $18 - 17 = 1$, and the number of half-filled orbitals is 1.

$\cdot CH_3$ has 7 valence electrons; the electron-count deficiency is therefore $8 - 7 = 1$, and the number of half-filled orbitals is 1.

SR_2 has 8 valence electrons; the electron-count deficiency is therefore $8 - 8 = 0$, and the number of half-filled orbitals is 0 (two lone pairs + two S—R bonds).

PR_3 has 8 valence electrons; the electron-count deficiency is therefore $8 - 8 = 0$, and the number of half-filled orbitals is 0 (three P—R bonds + one lone pair).

AsR_3 has 8 valence electrons; the electron-count deficiency is therefore $8 - 8 = 0$, and the number of half-filled orbitals is 0 (three As—R bonds + one lone pair).

Therefore

$$Ir(PR_3)_4 \longleftrightarrow \cdot CH_3 \text{ and } SR_2 \longleftrightarrow PR_3 \longleftrightarrow AsR_3 \text{ but}$$

$$Ir(PR_3)_4 \not\longleftrightarrow PR_3$$

(f) $Ni(CO)_2$ is an example of a d^{10} ML_2 fragment. The total number of electrons is $10 + 4 = 14$; the electron-count deficiency is therefore $18 - 14 = 4$, and the number of half-filled orbitals is 4.

$Os(CO)_3$ is an example of a d^8 ML_3 fragment. The total number of electrons is $8 + 6 = 14$; the electron-count deficiency is therefore $18 - 14 = 4$, and the number of half-filled orbitals is 4.

BH has 4 valence electrons; the electron-count deficiency is therefore $8 - 4 = 4$, and the number of half-filled orbitals is 4.

Therefore the three fragments are isolobal:

$$Ni(CO)_2 \longleftrightarrow Os(CO)_3 \longleftrightarrow BH$$

(g) As $Ni(CO)_2 \longleftrightarrow [BH]$ (part f), multiples of these fragments will also be isolobal; hence

$$[Ni_5(CO)_{10}] \longleftrightarrow [B_5H_5] \text{ and } [Ni_5(CO)_{10}]^{2-} \longleftrightarrow [B_5H_5]^{2-}$$

(h) As $Os(CO)_3 \longleftrightarrow BH$ (part f), it follows that

$$[Os_5(CO)_5] \longleftrightarrow [B_5H_5] \text{ and } [Os_5(CO)_{15}]^{2-} \longleftrightarrow [B_5H_5]^{2-}$$

SAQ 11
(*Objective 15*)

(a) $[Co_4(CO)_{12}]$ can be divided into four structural units—that is, four $Co(CO)_3$ units.

The number of electrons in each unit is given by $(V + X - 12) = 9 + 6 - 12 = 3$. This gives a total of $4 \times 3 = 12$ skeletal electrons, and thus six skeletal electron pairs.

This combination of four structural units and six electron pairs gives a *nido* structural-type classification and a trigonal bipyramidal shape with one vertex missing.

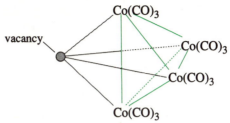

(b) $[Rh_6(CO)_{16}]$ can be divided into six structural units—that is, four $Rh(CO)_3$ units and two $Rh(CO)_2$ units.

For $Rh(CO)_3$ the number of electrons is given by $V + X - 12 = 9 + 6 - 12 = 3$, and for $Rh(CO)_2$ by $V + X - 12 = 9 + 4 - 12 = 1$. This gives a total of $4 \times Rh(CO)_3 = 12$ and $2 \times Rh(CO)_2 = 2$, or fourteen skeletal electrons—that is, seven electron pairs.

This combination of six structural units and seven electron pairs leads to a *closo* structural type classification. (This results in an octahedron or trigonal prism of metal atoms.)

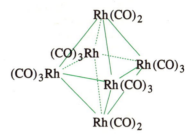

(c) $[Ni_5(CO)_{12}]^{2-}$ can be divided into five structural units—that is, two $Ni(CO)_3$ units and three $Ni(CO)_2$ units.

For $Ni(CO)_3$ the number of electrons is given by $V + X - 12 = 10 + 6 - 12 = 4$, and for $Ni(CO)_2$, $V + X - 12 = 10 + 4 - 12 = 2$. The total number of electrons is then given by $2 \times Ni(CO)_3 = 2 \times 4 = 8$, plus $3 \times Ni(CO)_2 = 3 \times 2 = 6$, plus the overall 2– charge (2 electrons); that is, $8 + 6 + 2 =$ sixteen skeletal electrons or eight electron pairs.

This combination of five structural units and eight electron pairs leads to an *arachno* structural-type classification and a pentagonal bipyramidal shape with two vertices missing.

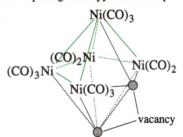

(d) $[Ru(CO)_3Co_2(CO)_6Se]$ can be divided into four structural units—that is, one $Ru(CO)_3$ unit, two $Co(CO)_3$ units and one Se unit.

For $Ru(CO)_3$ the number of electrons is given by $V + X - 12 = 8 + 6 - 12 = 2$; for $Co(CO)_3$, $V + X - 12 = 9 + 6 - 12 = 3$; and for Se, $V + X - 2 = 6 + 0 - 2 = 4$ (as selenium is a main-Group and not a transition element). The total number of electrons is then given by $2 + (2 \times 3) + 4 = 12$ electrons—that is, six electron pairs.

The combination of four structural units and six electron pairs leads to a *nido* structural-type classification. The shape of the molecule will be trigonal bipyramidal with one vertex missing.

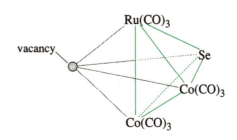

(e) [cpNiCo$_3$(CO)$_9$] can be divided into four structural units—that is, one cpNi unit and three Co(CO)$_3$ units. For cpNi the number of electrons is given by $V + X - 12 = 10 + 5 - 12 = 3$ (cp is a five-electron donor), and for Co(CO)$_3$, $V + X - 12 = 9 + 6 - 12 = 3$. The total number of skeletal electrons is then given by $3 + (3 \times 3) = 12$, or six electron pairs. The combination of four structural units and six electron pairs leads to a *nido* structural-type classification. The shape of the molecule will be trigonal bipyramidal with one vertex missing.

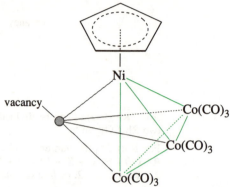

(f) B$_5$H$_9$ can be divided into five structural units—that is, four BH$_2$ units and one BH unit. For BH$_2$ the number of electrons is $V + X - 2 = 3 + 2 - 2 = 3$; for BH, $V + X - 2 = 3 + 1 - 2 = 2$. The total number of skeletal electrons is therefore $(4 \times 3) + 2 = 14$, or seven electron pairs. The combination of five structural units and seven electron pairs leads to a *nido* structural-type classification. The shape of the molecule will be based on an octahedron with one vertex missing.

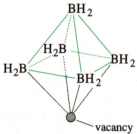

(g) C$_2$B$_4$H$_8$ can be divided into six structural units—that is, two carbons and four BH$_2$ units. The number of electrons in each structural unit is then given by: BH$_2$, $V + X - 2 = 3 + 2 - 2 = 3$; the two carbons donate two electrons each ($V + X - 2 = 4 + 0 - 2 = 2$). The total number of skeletal electrons is therefore $(4 \times 3) + (2 \times 2) = 16$, or eight electron pairs. The combination of six structural units and eight electron pairs leads to a *nido* structural-type classification. The shape of the molecule will be pentagonal bipyramidal with one vertex missing.

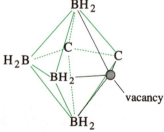

(h) [Ru$_2$Rh$_2$H$_2$(CO)$_{12}$] can be divided into four structural units—that is, two Rh(CO)$_3$ units and two RuH(CO)$_3$ units. For Rh(CO)$_3$ the number of electrons is given by $V + X - 12 = 9 + 6 - 12 = 3$, and for RuH(CO)$_3$, $V + X - 12 = 8 + 1 + 6 - 12 = 3$. The total number of skeletal electrons is therefore $(2 \times 3) + (2 \times 3) = 12$, or six electron pairs. The combination of four structural units and six electron pairs leads to a *nido* structural-type classification. The shape of the molecule will be trigonal bipyramidal with one vertex missing. This is confirmed by ^{1}H n.m.r., as indicated by the structure in Block 5, Figure 1.

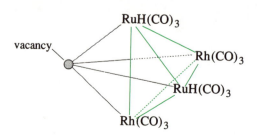

SAQ 12
(Objective 13)

(a) There will be no preference for the site of first attack. However, after the first addition has taken place, that ring will now be η^5-attached to the metal, as shown in the diagram. The second attack will be on the opposite ring, since the rule 'even before odd' will now apply.

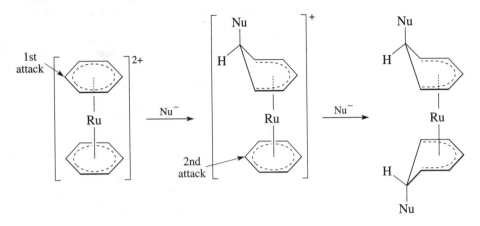

(b) Again there will be no preference for the site of first attack. In this case the intermediate formed after the first addition contains a ring that is η^4-bound to the metal, as shown in the diagram. This ring is now an even polyene, so the rule 'even before odd' means that the second attack will be on the same ring.

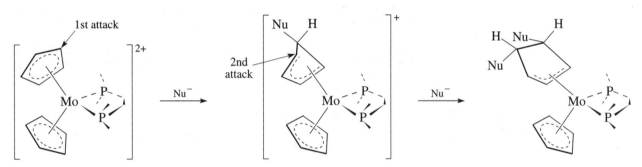

SAQ 13
(Objective 7)

The cyclohexadienyl group may bond through one, three or five carbon atoms to one or more metal atoms. Various ways in which the ring may bond to one metal atom are illustrated in Figure 69. The number of possibilities increases notably if the ring is attached to more than one metal; the additional bonding modes follow directly from those given in Figure 69.

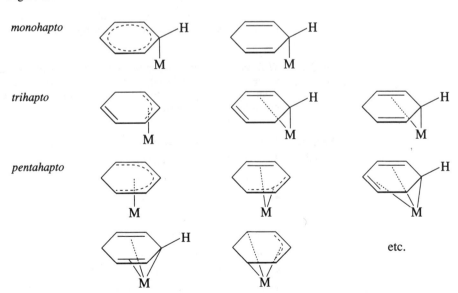

Figure 69 Some possible modes of attachment of a cyclohexadienyl group to one metal atom.

SAQ 14
(Objective 11)

(a) The total number of valence electrons in nickelocene, $[Ni(C_5H_5)_2]$, is $10 + (5 \times 2) = 20$.

(b) By reference to the molecular orbital energy-level diagram for ferrocene (Figure 47), it can be seen that the additional two electrons in nickelocene must be fed into the doubly degenerate $2\pi^*$ level. There will therefore be two unpaired electrons and the complex will be paramagnetic, with an expected magnetic moment of $2.83\mu_B$.

(c) By analogy with the synthetic methods used for other first-row transition-element metallocenes, an appropriate preparative route to nickelocene would be a metathetical reaction between a nickel dihalide and sodium cyclopentadienide:

$$NiX_2 + 2Na(C_5H_5) \longrightarrow [Ni(C_5H_5)_2] + 2NaX \qquad 119$$

(d) Many techniques could be used to show that nickelocene contains unsubstituted C_5H_5 rings. The two mentioned in the text are infrared spectroscopy, using the $1\,100\text{–}1\,000\,\text{cm}^{-1}$ rule (an unsubstituted ring has an absorption in this region), and mass spectroscopy, whereby the $Ni(C_5H_5)^+$ fragment would be prominent for unsubstituted species.

(e) Probable reactions of nickelocene include oxidation (reaction 87), protonation of the metal by a strong acid, transferal of the C_5H_5 group to another metal (reaction 91), metallation at the ring (as in Figure 54), and electrophilic aromatic substitution (Section 7.4.3).

SAQ 15
(Objective 22)

On formation of the metal–alkene bond, electron density is donated from the π-bonding orbital of the alkene to empty orbitals on the metal. This results in a decrease in the strength of the alkene C=C bond, as indicated by the lengthening of the bond and the decrease in the $\nu(C=C)$ stretching frequency. In addition, back-donation of electron density from the metal to the π^*-antibonding orbital of the alkene results in an increase in the electron density present in this antibonding orbital, with further weakening of the C=C bond.

Thus, both components of the metal–alkene bond unite to weaken the C=C bond. The lower the oxidation state of the metal, the greater is the degree of back-donation, and the less strongly are the d electrons held.

SAQ 16
(Objective 22)

The catalyst must fulfil the following functions:

(a) It must weaken or cleave the H—H bond in hydrogen. This is achieved in step **1** or step **9** of Figure 59.

(b) It must weaken the alkene C=C bond. This is achieved by coordination of the alkene as in step **3** or step **7** of Figure 59.

(c) It must hold the alkene and hydrogen atoms close together so that reaction between them can occur. (step **4** of Figure 59)

(d) It must not bind the hydride, alkene or alkane so strongly as to inhibit the reaction, so that products (alkanes) can readily leave the catalyst after reaction (step **6** of Figure 59).

SAQ 17
(Objectives 22c and 22d)

Wilkinson's catalyst depends on coordination of the alkene to a rhodium atom that also carries two bulky triphenylphosphine ligands. Very bulky alkenes may therefore be inhibited from a close approach to rhodium, and thus hydrogenation may not be possible in the presence of this catalyst.

SAQ 18
(Objectives 5, 19 and 21)

1 oxidative addition of hydrogen (hydrogenation);
2 loss of triphenylphosphine;
3 η^2-coordination of an alkene;
4 ligand combination reaction (insertion of alkene into a Rh—H bond);
5 addition of triphenylphosphine;
6 reductive elimination of an alkane;
7 η^2-coordination of an alkene;
8 loss of triphenylphosphine;
9 oxidative addition of hydrogen (hydrogenation).

Rhodium has sixteen valence electrons in $[RhCl(PPh_3)_3]$, $[Rh(CH_2CH_2R)ClH(PPh_3)_2]$, $[RhCl(H)_2(PPh_3)_2]$ and $[RhCl(RCH=CH_2)(PPh_3)_2]$. In all the other complexes in the catalytic cycle, rhodium has eighteen valence electrons.

Hence the intermediates in the Wilkinson catalytic cycle as drawn in Figure 59 do obey Tolman's rule.

SAQ 19 (Objective 19)

(a) The principal processes and requirements for homogeneous catalysis in solution are: closed-loop reaction cycle; few side reactions that would poison the catalyst; a reasonably fast catalytic reaction; mild conditions preferable.

The catalyst itself normally requires a metal with two stable oxidation states, and the catalytic cycle often involves reductive elimination and oxidative addition. Furthermore, the catalyst should not be poisoned by components in the cycle.

(b) Homogeneous catalysis has reactants and catalyst in the same phase (normally in solution); heterogeneous catalysis occurs at a phase interface (normally solid/gas or solid/liquid). Homogeneous catalysis is somewhat easier to model and reaction pathways are relatively easier to discern than in heterogeneous catalysis, although separation of the catalyst and products is easier for heterogeneous catalysis.

SAQ 20 (Objectives 5, 19 and 21)

(a) The oxidation states of rhodium in the intermediates are: **A**, +1; **B**, +3; **C**, +3; **D**, +3.

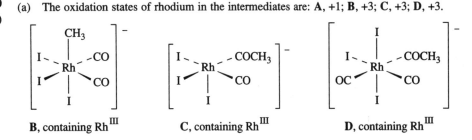

A, containing Rh^I **B**, containing Rh^{III} **C**, containing Rh^{III} **D**, containing Rh^{III}

(b) $[RhI_2(CO)_2]^-$ The number of valence electrons available to rhodium in this complex is 9 (Rh) + 2 ((I)$_2$) + 4 ((CO)$_2$) + 1 (negative charge) = 16 electrons

$[RhI_3(CO)_2(CH_3)]^-$ = 9 + 3 + 4 + 1 + 1 = 18 electrons

$[Rh(COCH_3)I_3(CO)]^-$ = 9 + 1 + 3 + 2 + 1 = 16 electrons

$[Rh(COCH_3)I_3(CO)_2]^-$ = 9 + 1 + 3 + 4 + 1 = 18 electrons

In other words all the intermediates obey Tolman's rule.

(c) step 1 oxidative addition;
step 2 ligand combination or insertion (CO inserts into the M—C bond);
step 3 addition (carbonylation);
step 4 reductive elimination (loss of CH_3COI).

(d) Two interdependent catalytic cycles are operating. The main cycle turns CH_3OH into CH_3COOH via carbonylation, and a subsidiary cycle converts CH_3COI to CH_3COOH and HI.

(e) $[RhI_2(CO)_2]^-$ and $[Rh(COCH_3)I_3(CO)]$ are the only coordinatively unsaturated species in the cycles.

SAQ 21 (Objectives 5, 19 and 21)

(a) 1 Alkene addition to a sixteen-electron rhodium species produces a complex in which the alkene is η^2-bound.
2 Ligand combination reaction (insertion of alkene into the Rh—H bond).
3 Addition of carbon monoxide.
4 Ligand combination reaction (insertion of carbon monoxide into the Rh—C bond).
5 Oxidative addition of hydrogen to give a dihydride.
6 Reductive elimination of an aldehyde.

(b)

	A			B	
	Rh	= 9		Rh	= 9
	H	= 1		$2 \times PPh_3$	= 4
	CO	= 2		CO	= 2
	$2 \times Ph_3P$	$= 2 \times 2 = 4$		CH_2CH_2R	= 1
	alkene	= 2			
	total	18 electrons		total	16 electrons

Thus, **A** obeys the eighteen-electron rule, but **B** does not.

(c) The rhodium–alkene bonding in compound **A** can be described as follows: σ donation from the alkene, together with synergic back-donation from the rhodium into the π* antibonding orbital of the alkene (Figure 70a). The rhodium–carbon monoxide bonding in compound **A** can be described as: σ donation from carbon monoxide and synergic back-donation from the rhodium to the CO π* orbital (Figure 70b).

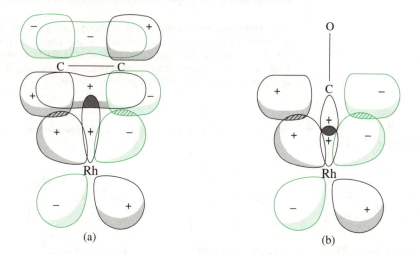

Figure 70 Metal–carbon bonding in compound **A** of the Union Carbide hydroformylation process: (a) orbital interaction between rhodium and the alkene; (b) orbital interaction between rhodium and carbon monoxide.

SAQ 22
(*Objectives 5, 21 and 22*)

(a) 1 coordination of η^2-bound butadiene;
2 ligand combination reaction (insertion of butadiene into the nickel–hydrogen bond);
3 reductive elimination of $CH_2=CHCH_2CH_2CN$;
4 oxidative addition of HCN to generate the active catalyst.

(b) $[Ni(CN)HL_2]$ is formed by ligand dissociation from NiL_4 to give NiL_2 followed by oxidative addition of HCN (L, such as PR_3, is a two-electron donor).

(c) NiL_2 is formally a fourteen-electron species, the others are sixteen- or eighteen-electron species. Tolman's rule states that sixteen- or eighteen-electron species are normal in organometallic reactions. The NiL_2 is unusual in that it is a fourteen-electron species.

S343 Inorganic Chemistry
Block 1   Introducing the transition elements
Block 2   Metal–ligand bonding
Block 3   Transition-metal chemistry: the stabilities of oxidation states
Block 4   Structure, geometry and synthesis of transition-metal complexes
Block 5   Nuclear magnetic resonance spectroscopy
Block 6   Organometallic chemistry
Block 7   Bioinorganic chemistry
Block 8   Solid-state chemistry
Block 9   Actinide chemistry and the nuclear fuel cycle